Fireground Size-Up
Study Guide

Fireground Size-Up
Study Guide

By Paul T. Dansbach
and Michael A. Terpak

Copyright© 2003 by
PennWell Corporation
1421 South Sheridan
Tulsa, Oklahoma 74112
800-752-9764
sales@pennwell.com
www.pennwell-store.com
www.pennwell.com

Supervising Editor: Jared d'orr Wicklund
Production Editor: Sue Rhodes Dodd
Cover Designer: Clark Bell
Layout Artist: Susan D. Kerr

ISBN 0-87814-897-3

All rights reserved. No part of this book may be reproduced, stored in a retrieval system, or transcribed in any form or by any means, electronic or mechanical, including photocopying and recording, without the prior written permission of the publisher.

Printed in the United States of America

1 2 3 4 5   07 06 05 04 03

This study guide is dedicated to all firefighters across our great nation who train to be better prepared to respond to the citizens of their communities in their time of need. As you use this study guide to hone your firefighting skills, remember the following:

- Training and education are investments in yourself and in your brother firefighters.

- The more you learn about the fire service, the more you realize there is always more to learn.

- Be heads up and be safe.

Contents

Foreword IX

Preface XI

Acknowledgments. XIII

Photo Credits XIII

1: The Fifteen Points of Size-Up
Chapter Questions 1

2: Private Dwellings
Chapter Questions 7
Photo Scenario Questions 13

3: Multiple Dwellings
Chapter Questions 17
Photo Scenario Questions 23

4: Taxpayers/Strip Malls and Stores
Chapter Questions 27
Photo Scenario Questions 33

5: Garden Apartments and Townhouses
Chapter Questions 39
Photo Scenario Questions 45

6: Row Frames and Brownstones
Chapter Questions 49

7: Churches
Chapter Questions 55
Photo Scenario Questions 61

Contents

8: Factories, Lofts, and Warehouses
Chapter Questions 65
Photo Scenario Questions 71

9: High-Rise
Chapter Questions 75
Photo Scenario Questions 81

10: Vacant Buildings
Chapter Questions 83
Photo Scenario Questions 89

Final Exam Questions 91

Foreword

When this project first began a number of years ago, the concept of size-up was limited in its definition and information. Definitions varied, information was vague, and others discredited its importance. It became quickly evident to me early on in my career as a firefighter in Jersey City that the information being shared, as well as personally viewed, often proved to be invaluable, and in many instances, it was lifesaving.

After years of observing and documenting the experiences of those who had been there before, and personally having the opportunity to work in a congested urban city, the text *Fireground Size-Up* was born (published in 2002 by PennWell).

It was my intent with this book that its content would be experienced, credible, useful, and easily referenced. It seems that those same objectives are being met again. Within the *Fireground Size-Up Study Guide*, Chief Paul Dansbach is carrying on this mission. A great friend and respected fire service educator, Paul furthers the information sharing by testing and evaluating the reader through use of multiple-choice questions and scenario-based responses on each of the occupancies discussed in the text. His approach in this companion study guide is direct and challenging. With size-up being the foundation for a fire officer's decision making, you should expect nothing less.

It will be a valuable addition to your fire service library.

Michael Terpak
Chief, 2nd Battalion
City of Jersey City Fire Department

Preface

Many of these chapters contain pictures of various occupancies and ask questions about points of size-up of the given building. The questions are intended to allow the student to apply the information in the text *Fireground Size-Up*. Understand that there may be many "correct" answers to the questions and we are limited by the view of the camera lens.

At the end of each chapter is an answer key showing the best answer to each question. Also included are references to the pages in Michael Terpak's *Fireground Size-Up* text to aid the reader in further study.

Fire instructors, chief officers, and company officers are encouraged to use this training format in their local departments by using photos and images of buildings. This can be a very valuable training tool for your fire department as you reinforce and test your members' knowledge while gaining important information and a better understanding of the buildings in which you may be called upon to fight fire.

Never stop training . . . Never, Never, Never, Never!

Paul Dansbach

Acknowledgments

I would like to thank Michael A. Terpak for the opportunity to develop and write this study guide. Mike's confidence in my ability, encouragement in this project, and words of wisdom: "Don't make the questions too hard" were invaluable while preparing this text. Mike is truly a great fire service instructor, a natural fire service leader—and above all, Mike is a great friend. Thanks Mike!

Photo Credits

Paul T. Dansbach
Courtney A. Dansbach
Christian P. Dansbach
Michael A. Terpak — Jersey City, New Jersey Fire Department
Roy Van De Voort — Haledon, New Jersey Fire Department
Erik Valez — Wood-Ridge, New Jersey Fire Department
Anthony Greco — Hasbrouck Heights, New Jersey Fire Department,
    FDIC Photographer

1

The Fifteen Points of Size-Up

Chapter Questions

1) **Alterations and renovations to buildings have resulted in hybrid construction. It is important for the fire officer to recognize hybrid construction because:**

    a) Hybrid construction always uses steel to replace wood structural elements.
    b) Hybrid construction may use lightweight structural elements, thereby reducing the collapse resistiveness of certain buildings.
    c) Hybrid construction is easy to recognize, so preplanning of these structures is not important.
    d) Hybrid construction uses "engineered structural elements" that have the same characteristics as the original structural components they are replacing.

2) **The term area is defined as the square footage involved in or threatened by the fire. The following building features should be identified to determine the area of a building:**

    a) Basement level area and structural support system.
    b) 1st-floor level area and access to the rear of the building.
    c) Irregular-shaped buildings and interconnected buildings.
    d) Mezzanine areas and access to the mezzanine level.

Fireground Size-Up Study Guide

3) The following are occupant life hazard concerns:

   a) Time of day and the size of the building.
   b) Street conditions for apparatus access and terrain for ground ladder access.
   c) The number of and location of the occupants.
   d) Resources necessary and type of construction.

4) Which of the following occupancy characteristics may have an impact on the structural integrity of the building?

   a) High-value occupancies, such as museums.
   b) Buildings containing water-absorbent materials.
   c) Assembly occupancies, such as nightclubs.
   d) Research facilities employing poisonous and biohazard materials.

5) The **Life Hazard** size-up is without question the most important size-up point to a fire officer. Which of the following size-up points influence the decision-making process in determining the **Life Hazard** size-up?

   a) Construction, occupancy, and street conditions.
   b) Occupancy, height, and time of day.
   c) Street conditions, occupancy, and time of day.
   d) Occupancy, location and extent of fire, and time of day.

6) **Location** and **Extent** of fire size-up point categorize the fire location into four categories. Each separate category identifies specific extension potentials, resource requirements, life hazard potential, and ventilation problems. Choose the four correct categories:

   a) Fires in stairways, fires below grade, fires in shafts, and top floor fires.
   b) Fires below grade, fires at lower levels of the structure, top floor and attic/cockloft fires, and fires beyond the reach of fire department aerial apparatus.
   c) Fires in shafts, fires below grade, fires beyond the reach of fire department aerial apparatus.
   d) Fires in shafts, fires below grade, fires on floors accessible by fire department aerial apparatus, and fires beyond the reach of fire department aerial apparatus.

7) The size-up point **Street Conditions** include the following considerations:

   a) Street width, terrain, and street surface.
   b) Terrain, traffic flow, and unusual condition such as flooding.
   c) Street width, traffic flow, and street surface.
   d) Terrain, traffic flow, and weather.

8) When fighting a fire in a cockloft, introduction of air into a concealed space from below may:

   a) Cause the ceiling to explode down, trapping firefighters.
   b) Cause the ceiling area to flashover, exposing firefighters to extreme temperatures.
   c) Have no effect on the fire in the cockloft.
   d) Both **a** and **b** are correct answers.

9) Which are the three primary correct choices for size-up considerations for below-grade fires?

   a) Time of day, access to the below-grade level, and ventilation.
   b) Time of day, fire loading, and access to the below-grade level.
   c) Access to the below-grade level, ventilation, and day of the week.
   d) Access to the below-grade level, ventilation, and fire loading.

10) For the size-up factor **Weather**, which two factors have the greatest impact on firefighter safety?

    a) Wind and temperature.
    b) Temperature and humidity.
    c) Humidity and precipitation.
    d) Precipitation and wind.

Fireground Size-Up Study Guide

11) **Specialized extinguishing equipment includes systems capable of extinguishing a specific type of fire within a given occupancy. These systems include which of the following?**

   a) Fixed wet or dry chemical systems and dry pipe sprinkler systems.
   b) Fixed foam systems, fixed wet or dry chemical systems, and wet pipe sprinkler systems.
   c) Fixed foam systems, wet or dry chemical systems, and halon or other clean-agent systems.
   d) Fixed wet or dry chemical systems, halon or other clean-agent systems, and wet pipe automatic sprinkler systems.

12) **The water supply required for a specific incident will be determined by which of the following common factors?**

   a) Location and extent of the fire, time of day, construction class and fire loading, and fixed fire protection systems in the building.
   b) Location and extent of the fire, height and area of the fire building, class of construction, and contents/fire loading of the building.
   c) Location and extent of the fire, height and area of the building, class of construction, and street condition or accessibility to the structure.
   d) Location and extent of the fire, apparatus and staffing responding to the incident, class of construction, and contents/fire loading of the building.

13) **Buildings built on a sloped grade cause an increased life hazard concern to firefighters because:**

   a) A building that is 2 stories in the front may, in fact, be 3 or 4 stories from the rear; this may cause confusion and make fireground management more difficult.
   b) Steep slopes to grade will always preclude deployment of ground ladders to upper stories, placing firefighters searching the upper stories of the building at risk.
   c) Stories that open to a different grade at the rear of the building may confuse hoseline deployment by confusing which level is the 1st floor of the structure; this will make fireground management more difficult.
   d) Steep slopes to grade will make the use of aerial ladders more difficult and dangerous to firefighters.

14) The categories in the size-up point **Terrain** include which of the following:

   a) Setbacks, street width, and buildings built on a grade.
   b) Street width, buildings built on a grade, and setbacks.
   c) General accessibility, street width, and buildings built on a grade.
   d) General accessibility, buildings built on a grade, and setbacks.

15) Which of the following will influence initial occupant life hazard considerations within a fire building?

   a) The location and extent of the fire and access to the fire building.
   b) The location of the trapped occupants and access to the fire building.
   c) The location and extent of the fire and the buildings' areas of greatest danger.
   d) The location of the trapped occupants and placement of the first deployed hoseline.

Answer Key

Question #	Answer	Page #
1	b	1
2	c	41
3	c	24
4	b	18
5	d	22
6	b	42, 43
7	c	33, 34
8	a	45
9	d	43
10	b	35, 38
11	c	32
12	b	28
13	a	26
14	d	25
15	c	23

2 Private Dwellings

CHAPTER QUESTIONS

1) **Queen Anne/Victorian Style structures are examples of wood-frame construction. These structures include some of the following features:**

   a) They are 2½ to 3½ stories high, typically balloon-frame construction with a hip roof.
   b) They are 1½ to 2½ stories high, typically balloon-frame construction with a roof containing many peaks, valleys, and dormers.
   c) They are 1½ to 2½ stories high, typically balloon-frame construction with large eaves/overhangs and a cupola on the roof.
   d) They are 2½ to 3½ stories high, typically balloon-frame construction with a roof containing many peaks, valleys, and dormers.

2) **In wood-frame private dwellings, hazards of hybrid construction include:**

   a) The use of lightweight floor and roof trusses and unprotected structural steel components.
   b) The failure of gusset plate connectors and the use of recycled structural components.
   c) The use of unprotected structural steel to support concrete floors and poured over wood joists.
   d) The use of lightweight roof trusses with flake board on roof assemblies.

3) **Typically in wood-frame 2-family dwellings with apartments on each floor, the bedrooms are found at the following locations:**

   a) Always adjacent to the kitchen.
   b) Along the side of the building opposite the entry doors on the front of the house.
   c) In the rear of the building.
   d) There is no typical location for bedrooms in 2-family structures.

4) **Energy efficient windows are typically found in all new and renovated private dwellings. These windows affect fireground operations because they:**

   a) Retard fire growth, flashover never occurs, and ventilation is made more difficult.
   b) Retard fire growth, make locating the fire difficult, and allow backdraft conditions to occur.
   c) Are usually smaller in size, and ventilation will always be required to locate the seat of the fire.
   d) Have stronger window panes, making ventilation more difficult and creating more frequent flashover conditions.

5) **Which of the following are indicators that the attic area or level may be occupied?**

   a) Oversized windows on this level, multiple cars in the driveway, and TV antennas/satellite dishes attached at this level.
   b) Oversized windows on this level, air conditioners in attic windows, and skylights in the attic areas.
   c) Many windows in the attic, separate electric service for the attic, and multiple cars in the driveway.
   d) Fire escapes on the rear of the building, multiple cars in the driveway, and several names on the mailbox.

6) **Collapse considerations of private dwellings include the following:**

   a) Type of framing system (balloon or platform), exposure problems, and terrain.
   b) The structure is made entirely of wood, exposure problems, and age of the structure.
   c) The structure is made entirely of wood, hybrid materials may be used in the structure, and the lack of mass (therefore collapse resistiveness).
   d) Type of framing (balloon or platform), access to the rear of the structure, and age of the structure.

Private Dwellings

7) **The most important size-up factor in considering occupants Life Hazard is:**

   a) Time of the year.
   b) Day of the week.
   c) Time of the day.
   d) Occupancy classification.

8) **Hoseline selection for fires in private dwellings will be based on the following considerations:**

   a) Relatively light fire load, location and extent of fire, and small room sizes.
   b) Relatively light fire load and location and extent of fire extension probability.
   c) Relatively light fire load, small room size, and the need for a quick, mobile flexible hoseline.
   d) Relatively light fire load, small room size, and below-grade fires.

9) **What factors should a fire officer take into consideration for fires in private dwellings that are set back a significant distance from the street?**

   a) Can the engine company access the driveway and hoseline selection due to increased friction loss?
   b) Hoseline selection due to increased friction loss and the speed at which the back-up line can be stretched and placed in service.
   c) Hoseline selection due to increased friction loss and the location of the nearest hydrant or drafting location.
   d) The location of the nearest hydrant or drafting location and can the engine company access the driveway.

10) **The typical construction of wood-frame buildings can be classified as:**

    a) Class 5 — newer buildings of platform-frame construction.
    b) Class 3 — construction using steel as an alternative material where large wood beams were once used.
    c) Class 5 — an increase in the use of lightweight structural components creating hybrid types of construction.
    d) Class 3 — an increase in the use of lightweight structural components creating hybrid types of construction.

Fireground Size-Up Study Guide

11) **Which of the following factors should be considered when establishing a water supply at fires in private dwellings?**

   a) The 1st arriving company should always establish a water supply if the fire hydrant is within 500 ft of the fire building.
   b) Both the 2nd and 3rd due engine companies should establish water supplies.
   c) A quick and sustained water supply is a necessary part of successfully controlling fires in private dwellings.
   d) The 1st engine company should get a line in service to knock down the fire and protect the firefighters searching the building. Water supply will be established by the 2nd due engine company.

12) **Queen Anne or Victorian-type private dwellings are most commonly constructed of what?**

   a) Class 5 — platform frame.
   b) Class 5 — balloon frame.
   c) Class 5 — hybrid construction.
   d) Both **b** & **c** are correct answers.

13) **Vertical ventilation or roof ventilation of fires in private dwellings should be performed when:**

   a) The fire is in the basement of a 2½-story platform-frame structure.
   b) The fire is in the basement of a 2½-story balloon-frame structure.
   c) The fire is on the 1st floor and may have extended to the framing of a 2½-story balloon-frame structure.
   d) Both **b** & **c** are correct answers.

Private Dwellings

14) When considering the "area" of fire involvement, fires in larger private dwellings with high ceilings result in _____ while smaller private dwellings with smaller rooms and lower ceilings are more likely to _____.

   a) More concentrated area at the fire involvement ... have the fire spread to the next story or the attic of the private dwelling.
   b) More concentrated area of fire involvement ... have the fire spread beyond the room of origin.
   c) A slower developing fire with a great potential for backdraft ... have a rapid developing fire with a great potential for flashover.
   d) A slower developing fire with a great potential for flashover ... have a rapid developing fire with a great potential for backdraft.

15) Which of the following statements is true regarding the unenclosed interior stairway in a private dwelling?

   a) The interior stairway is not a major concern as most fires start on the 2nd or 3rd floor of the private dwelling.
   b) This stairway will allow unrestricted movement of fire and smoke immediately to the upper floors of the private dwelling.
   c) The backup hoseline must always be stretched to the upper floors of private dwellings to prevent fire extension to the upper floors.
   d) The interior stairway is only a major concern when the fire is located in the basement or below-grade levels.

Answer Key

Question #	Answer	Page #
1	d	55
2	a	52
3	c	63
4	b	61, 64
5	b	57
6	c	56, 57
7	c	67
8	c	59
9	b	70
10	c	51
11	c	58
12	b	62
13	d	62
14	c	78
15	b	79

Photo Scenario Questions

Photo 2–1

Based on Photo 2–1, answer the following questions:

1. What is the most likely type of construction for this building?

2. What will be the most likely path for vertical fire extension?

3. What will be the next most likely path for vertical fire extension?

4. From the picture, can you identify where the interior stairway is in the building?

5. The exterior door on the "D" side will provide ready access to what?

Answers

1. Balloon frame.

2. The interior stairway.

3. The balloon-frame interior and exterior walls.

4. Along the "D" exterior wall, the indicators of this are the small windows located most likely at a stair landing and the exterior door on the "D" side. This door opens into the stair landing between the basement and the 1st floor; the stairway between the 1st floor and the 2nd floor is typically stacked over the basement stairway.

5. Conduct a quick primary search of the basement and secure the electric and gas utilities. Crews should also monitor this area to ensure burning embers do not drop down through the balloon-frame partitions and start secondary fires in the basement.

COAL TWAS WEALTHS

Photo 2–2

Based on Photo 2–2, if you were the first arriving chief officer, what size-up points would you use to establish your incident priorities?

Answers

Construction — wood-frame construction. The entire building is combustible and likely to be balloon-frame construction.

Occupancy, Time and **Life Hazard** — residential 2-family dwelling. The fire is occurring during the day when the life hazard is low for this type of occupancy. Occupants of the exposures will likely self-evacuate.

Apparatus and **Staffing** — additional firefighting resources will be required. The fire building is well involved and both the "B" and "D" exposures are severely threatened. The amount and type of resources will depend on your department's initial response and resources available through your mutual aid agreements.

Water Supply — there is a large volume of fire on arrival, and the exposures are severely threatened. Additional large volumes of water will be necessary. Additional resources may be required to establish adequate water supplies.

Street Condition — is the street wide enough to accommodate several aerial companies? Will the engine companies have access to drop supply lines?

Location and **Extent of Fire** — the fire building is well involved and threatening both the "B" and "D" exposures.

Exposures — both exposures are threatened and are of combustible construction and both have many openings in the exterior walls.

Weather — the wind is impacting this fire. There is potential spread of the fire to the "D" exposure building.

3 Multiple Dwellings

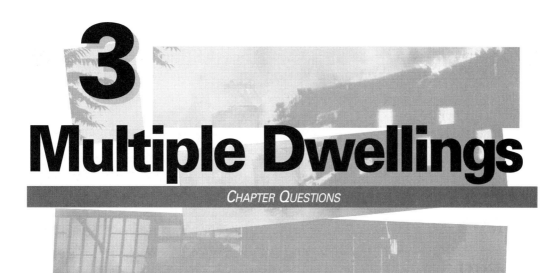

Chapter Questions

1) If a fire extends into a light and air shaft of a multiple dwelling, the following consideration is important:

   a) It is good if the fire extends to the light and air shaft as this will provide ventilation for the advancing engine company.
   b) You cannot identify fire extension into the light and air shaft from the street.
   c) When the fire extends to the light and air shaft, all apartments with openings into the shaft will be exposed.
   d) Fire extension into the light and air shaft will have little impact on the firefighting strategy.

2) Multiple dwellings of hybrid construction have the following construction features:

   a) Exterior masonry walls, wood joist floors, and bar joist roof assemblies.
   b) Exterior masonry walls, wood joist floors, and a heavy timber truss roof supported by steel columns.
   c) Exterior masonry walls, wood joist floors, wood rafters, and Type 1 spreader to support the exterior walls.
   d) Exterior masonry walls, steel beams or girders supporting wood floor joists, steel columns boxed out with plaster, and wood lath.

Fireground Size-Up Study Guide

3) **When fire escapes are located on the front and rear of a tenement-type multiple dwelling, this is an indication that:**

   a) There are only 2 apartments on each floor.
   b) One fire escape was added sometime after the building was constructed.
   c) There are 4 or more apartments on each floor.
   d) The number and location of fire escapes has no direct correlation to the number of apartments on each floor.

4) **When responding to a fire in any multiple dwelling, the main strategic factor is _____, and we must ensure our response to such fires have adequate _____.**

   a) The time of day ... staffing.
   b) Extension probability ... rescue companies.
   c) The life hazard ... staffing.
   d) Location of the fire ... EMS units.

5) **Of the following choices, which type of multiple dwellings present the greatest safety concern to firefighters?**

   a) High-rise buildings.
   b) Old wood-frame tenements.
   c) Non-fireproof buildings 4 to 6 stories high.
   d) Condominium and townhouse-type structures of lightweight frame construction.

6) **The occupant life hazard profile for fires in multiple dwellings must follow the assigning of priority areas based on the areas of greatest danger within the building. Those areas are:**

   a) Fire floor, areas connected on the same level of the fire floor, and the floor above the fire.
   b) Fire floor, floor above the fire, and floor below the fire.
   c) Fire floor, floor above the fire, and the next floor above the fire.
   d) Fire floor, floor above the fire, and the top floor of the building.

7) *What must be an automatic thought with a fire in a tenement or non-fireproof multiple dwelling is:*

   a) Ventilation of the fire room or apartment.
   b) Positive pressure ventilation behind the advancing engine company.
   c) Ventilation of the cockloft.
   d) Ventilation of the interior stairs.

8) *An important safety consideration when searching the apartment above the fire is:*

   a) Before entering the apartment, force another door on this floor, preferably the door directly across the hallway from the apartment to provide an area of refuge.
   b) Always search the apartment immediately above the fire with a charged hoseline after the apartment has been vented. This will enable the crew to ensure the area stays tenable.
   c) One firefighter should have a personal safety rope and the 2nd firefighter should have a lifeline. This equipment will ensure the crew can find their way back to the public hallway or bail out a window if conditions deteriorate.
   d) The ladder company should ventilate windows of the apartment directly above the fire and leave a ladder at a window as a 2nd means of escape for the search crew. A hoseline should also be stretched to the top of the ladder to ensure the room or apartment stays tenable.

9) *Firefighters who force the bulkhead door from the roof should always do the following:*

   a) Leave the roof immediately.
   b) Remain at the bulkhead door to monitor the ventilation progress and to ensure the door does not blow closed.
   c) Search the landing and stairway inside the bulkhead door.
   d) Remove the door from the hinges to ensure it does not blow closed.

10) **Which answer establishes priorities for fires in basements or below-grade areas in multiple dwellings with an open interior stairway?**

   a) Get a hoseline to the seat of the fire and vent the fire area.
   b) Use an exterior stream to knockdown the main body of fire and provide ventilation at the top of the stairway.
   c) Place a hoseline in service to protect the stairway and provide ventilation at the top of the stairway.
   d) Place a hoseline in service to protect the stairway and ventilate the fire area and the 1st floor.

11) **Because many multiple dwellings are large and irregularly shaped, it is important to report the following conditions from the rear or sides of the building to the operations officer or incident commander:**

   a) Location of person(s) trapped, location and extent of the fire, and access for firefighters and apparatus.
   b) Location of person(s) trapped, location and extent of the fire, and number of fire escapes.
   c) Location of person(s) trapped, exposure buildings, and hydrant locations.
   d) Location of person(s) trapped, access to fire escapes, and exposure buildings.

12) **The time of day plays a major role in fires occurring in multiple dwellings. The hours of midnight to 7 A.M. are the most difficult times for fires in these buildings because:**

   a) It is always more difficult to fight a fire at night.
   b) The building is most populated during these hours and many of the occupants are sleeping.
   c) Arson fires usually occur during these hours.
   d) There is usually a delay in reporting fires during these hours.

Multiple Dwellings

13) The largest concealed space in an apartment building or H-type multiple dwelling is the_____.

   a) Crawl space not considered a basement.
   b) Old dumbwaiter shafts that have been abandoned.
   c) Cockloft space.
   d) Void spaces of floor ceiling assemblies when constructed of lightweight wood trusses.

14) Light and air shafts will be found in two different design types. They are:

   a) Covered and uncovered, referring to a covering being in place on the top of the shaft at the roof level.
   b) Full height or partial height, meaning full height shafts extend from the basement through the roof while partial height shafts extend from the 1st floor to the roof.
   c) Combustible or non-combustible shafts, referring to the siding applied to the walls of the shaft.
   d) Open and enclosed shafts, with the open shaft being open to the rear yard and the enclosed shaft being surrounded by walls on all four sides.

15) Which of the following statements is true regarding the occupant life hazard of fires in Class 1 — fire-resistive multiple dwellings?

   a) All occupants from the fire floor through the top floor should be removed from the building as soon as resources permit.
   b) Occupants need only be removed from the fire floor, the floor directly above the fire, and the top floor of the building.
   c) Depending on the severity of the fire, you may leave the occupants in place in nearby apartments as conditions are often safer and more tenable in their apartments.
   d) Depending on the severity of the fire, you may leave non-ambulatory occupants in place in nearby apartments. A minimum of 2 firefighters equipped with a portable radio should stay with the occupant to monitor conditions in the apartment.

21

Answer Key

Question #	Answer	Page #
1	c	93
2	d	100
3	c	100
4	c	101
5	b	109
6	d	113
7	d	105
8	a	106
9	c	106
10	c	130
11	a	140
12	b	140
13	c	133
14	d	124
15	c	112

Multiple Dwellings

PHOTO SCENARIO QUESTIONS

Photo 3–1

Photo 3–2

Photo 3–3

COAL TWAS WEALTHS

If you were the first arriving fire officer at this structure (Photo 3–2) at 2 A.M., and fire was showing from the 1st floor windows as indicated by the arrows in Photos 3–1 and 3–3, what size-up points would influence your firefighting strategy and tactics?

23

Answers

Life Hazard, Occupancy, and **Time** — residential 6-family occupancy and the fire is occurring at 2 A.M. This would indicate a severe life hazard as the occupants are most likely sleeping.

Location and **Extent of Fire** — fire is showing out 2 windows on the 1st floor indicating the fire is well developed in one apartment. The most serious fire extension problem is the interior stairway. Maintaining control of the interior stairway will be essential for the evacuation of the occupants and successful control of the fire.

Construction — the building is wood-frame balloon construction. The structure is combustible and the balloon frame has potential for vertical fire extension.

Apparatus and **Staffing** — search and rescue operations will be very labor intensive and will likely require additional resources. The amount and type of additional resources required will depend on your department's initial resources as well as those available in your mutual aid.

Height and **Area** — the building's square footage is not very large and the building is 3 stories high. These points do not pose any unusual problems.

Street Conditions — the building is located at a street intersection. There are few if any overhead obstructions.

Terrain — this size-up point poses no unusual problems. Access to the fire escapes on the "C" and "D" sides of the building is easy. The fire escapes provide ready access for search, rescue, and ventilation. The rear fire escape provides access to the roof.

Multiple Dwellings

Photo 3–4

Photo 3–5

Based on Photos 3–4 and 3–5, answer the following questions.

1. What is the construction type for this building?

2. What is the occupancy of this building?

3. Based on the views of this building, can you determine how many apartments there are per floor in this building?

4. What size-up points does the number of apartments impact?

5. How does the size-up point **Street Conditions** affect your fireground operations?

6. How will the size-up point **Terrain** affect our fireground operations?

ANSWERS

1. The building is ordinary type construction.

2. The building contains apartments above stores on the 1st floor.

3. A rear porch contains a stairway, and fire escapes are located on the front of the building. The entrance to the apartments is located in the center of the building. These are indicators that the building will have at least 4 apartments per floor.

4. **Life Hazard, Apparatus and Staffing, and Area.**

5. The width of the street appears to be satisfactory.

6. Access to the rear of the building must be made through or around adjoining structures or from the street at the back of the building. There is adequate open space in the parking lot on the "B" side to set up aerial devices. There is also a utility pole with wires running across the front of the building.

4
Taxpayers/Strip Malls and Stores

Chapter Questions

1) **The major difference between a strip mall and a taxpayer is:**

   a) Urban vs. suburban setting.
   b) Age and construction type.
   c) Setback and access to the rear of the structure.
   d) The number of stores and apartments on the floors above.

2) **The floors of many taxpayer-type buildings are constructed of terrazzo, marble, or poured concrete. The fire services' main concern with this type of floor is:**

   a) The floor allows water to pool on the floor.
   b) These materials are often disguised when covered with tile or carpet.
   c) There may be little indication of the fire below, and a weakened floor assembly may go unnoticed.
   d) These materials are very difficult to cut in order to open the floor for ventilation purposes.

3) **Below-grade fires in taxpayers and stores are difficult and dangerous for which of the following reasons?**

   a) Low ceiling heights, lack of sprinkler systems, and poor accessibility to these areas.
   b) Limited access, heavy fire loading, and maze-like conditions.
   c) Low ceiling heights, ventilation is impossible, and heavy fire loading.
   d) Limited access, fire extension to the 1st floor is likely, and ventilation is difficult.

4) Fires occurring in the 1st floor rear storage area in strip malls and taxpayers often result in:

   a) Delayed alarm and advanced fire conditions on arrival.
   b) Accessibility problems since the rear door is well secured.
   c) Delayed alarm and fire extension to the floors above.
   d) Fire extension into the drop ceiling in the customer area of the occupancy and the cockloft space.

5) The probability of fire extension in taxpayers from the original fire occupancy to the exposure occupancies is high for the following reasons:

   a) Delayed alarms lead to advanced fire conditions on arrival.
   b) Common cockloft and high fire loads associated with these occupancies.
   c) Roofs on these structures are very difficult to effectively ventilate.
   d) Firewalls are never present in this type of occupancy.

6) Firefighters advancing a hoseline into a store or strip mall should consider the following hazards:

   a) Construction type, high-piled stock, and maze-like arrangement.
   b) Large floor areas, high-piled stock, and maze-like arrangement.
   c) Construction type, high-piled stock, and large floor areas.
   d) Large floor areas, poor access into the rear of the structure, and maze-like arrangements.

7) Which type of roof system will negate rooftop ventilation at a taxpayer or strip mall?

   a) Metal deck.
   b) Wood joist.
   c) Heavy timber.
   d) Lightweight wood truss.

Taxpayers/Strip Malls and Stores

8) *When a fire vents out the large store front windows of a 1-story taxpayer, the parapet wall may be at risk of collapse because:*

   a) The pressure differential created when the fire vents out the large window opening at the front of the building.
   b) The steel beam used to create the window opening may twist as the result of heating and drop the parapet wall into the street.
   c) The wood roof assembly may burn through immediately behind the large opening, leaving the parapet wall with little or no support.
   d) Parapet walls are always in danger of collapse because they are unsupported at the ends.

9) *When no specific incident information is provided or available, arriving officers may gather occupancy hazard information from:*

   a) The owner or manager of the business when they arrive.
   b) The owner or manager of the adjoining business who may be present.
   c) The occupancy sign on the front of the building.
   d) Nothing. Without accurate preplan information, there is no good way to gather information regarding the occupancy hazard.

10) *Modern taxpayer or strip mall buildings are typically constructed of Class 2 — non-combustible/limited combustible construction. The roof system on these structures is typically:*

    a) Metal bar joist.
    b) Lightweight concrete plank.
    c) Lightweight wood truss.
    d) Both **a** and **c** are correct answers.

11) *Taxpayer/strip malls and stores will most often have a high fire loading and large, deep floor areas. Which of the following are considerations in hoseline selection?*

    a) Hoseline mobility, stream volume (GPM), and stream reach.
    b) Stream volume (GPM), stream reach, and stream penetration.
    c) Stream reach, stream penetration, and hoseline mobility.
    d) Stream penetration, stream volume (GPM), and available water supply.

Fireground Size-Up Study Guide

12) **According to annual statistics compiled by the National Fire Protection Association (NFPA), firefighters are _____ times more likely to die in a commercial building fire than they are in a residential building fire.**

   a) 1½ times.
   b) 2 times.
   c) 2½ times.
   d) 3 times.

13) **Which of the following conditions or considerations will directly impact firefighter safety?**

   a) Familiarity with the structure, advanced fire conditions, contents and square footage, firefighter disorientation, and collapse.
   b) Familiarity with the structure, advanced fire conditions, construction type, extension probability, and time of day.
   c) Familiarity with the structure, advanced fire conditions, time of the year, extension probability, and number of stories/height of the building.
   d) Both **a** and **c** are correct answers.

14) **Time of day is a factor for fires in commercial occupancies because of the following:**

   a) The life hazard decreases as these occupancies are never occupied when closed for business.
   b) When a fire occurs during business hours, the fire is reported promptly and the occupants will self-evacuate.
   c) Security measures such as roll-down steel doors and security gates will result in delayed forced entry, ventilation, and water application on the fire.
   d) Time of day is not a consideration or concern to fire officers responding to or operating at fires in commercial occupancies.

15) *Occupancies such as meat markets and butcher shops have insulated ceilings over the frozen storage areas. These ceilings contain plywood nailed or screwed to the ceiling joists followed by sheets of insulating material. Fire in the cockloft space above these areas may require unconventional means to extinguish the fire in this area. Which unconventional means may be employed to best extinguish the fire in this area?*

   a) Direct water through the roof into this area.
   b) Remove the ceilings entirely from the perimeter of this area, then firefighters can climb to the top of this area using a pencil ladder.
   c) Cut the ceiling away in strips, cutting parallel to the ceiling joists to expose this area.
   d) Both **b** and **c** are correct answers.

Answer Key

Question #	Answer	Page #
1	b	144
2	c	145, 146
3	b	168
4	d	170
5	b	164
6	b	157
7	d	154
8	b	146
9	c	149
10	d	147
11	b	152
12	b	156
13	a	156, 157
14	c	172
15	a	171

Taxpayers/Strip Malls and Stores

PHOTO SCENARIO QUESTIONS

Photo 4–1

COAL TWAS WEALTHS

You are the first arriving company officer and smoke is pushing from the top of the roll-down security gates in the occupancy as indicated by the arrow. Based on this photo, what size-up points will you use to establish your incident priorities?

Answers

Construction — the building is an ordinary construction structure and most likely will have a common cockloft.

Occupancy — the building houses stores or mercantile occupancies. The fire load may be significant and access through the store may be limited by row of stock and display racks.

Apparatus and **Staffing** — resources will be required to force open all the roll-down security gates starting with the occupancy with the heaviest smoke condition. If your response is 3 engine companies and 1 truck company, additional resources will be necessary because of the number of security gates present. Keep in mind that additional truck company functions will also be necessary. For example: forcing doors at the rear of the building, ventilation of the roof, and searching the occupancy. A minimum additional truck company and a rescue company should be called because the task of opening the security door will require additional staffing and tools.

Life Hazard — the stores all appear to be closed, therefore, the occupant life hazard concern is low. Our other life hazard concern is the safety of the firefighters. Operating in stores will pose certain hazards such as a large fire load, narrow aisles, and limited access to the inside of the structure.

Weather — wind direction and speed will be a consideration if the fire extends to the cockloft area.

Area — the building area does not pose any significant problems. Companies must be assigned to operate at the rear of the building; these companies should report conditions to the incident commander. Their report should include, but not be limited to, the following: fire condition, access into the building, obvious fire walls, irregular shaped buildings, and any additional stories such as a basement or 2nd floor, which may not be obvious from the front of the building. A report from the roof will help identify any fire walls that might be present separating the occupancies and help confine the fire.

Taxpayers/Strip Malls and Stores

Photo 4–2

Photo 4–3

Based on Photos 4–2 and 4–3, answer the following questions.

1. What is the construction type for this building?

2. What is the occupancy of the building?

3. What are your life hazard concerns for the building?

4. Where is the entrance to the residential apartments?

5. How does the layout and arrangement of the building affect fireground management?

Answers

1. The front 1-story area of the building appears to be ordinary construction. The 3-story rear area of the building appears to be balloon-frame wood construction.

2. The 1st floor of the front area of the building is occupied as stores. The rear 2nd- and 3rd-floor areas are residential apartments.

3. The most significant life hazard will be the residential apartments at the rear of the building. The most significant life threat will be between 11 P.M. and 7 A.M. while the residents are likely to be sleeping. Our other life hazard concern is the safety of the firefighters. Operating in stores will pose certain hazards such as a large fire load, narrow aisles, and limited access to the inside of the structure.

4. The rear or the "C" side of the building.

5. This building has two separate entrances. The front of the building on the main street has the entrances to the stores located on the 1st floor. The entrances to the apartments are located at the rear of the building, which is accessible from the street to the rear of the building. Standing in the front of the building, your line of sight to the rear 3-story residential area is limited. The area of the building is unique in this respect. It will be important to get a report from the rear of the building early on in this incident. If the fire is in the 1st floor occupancies, fireground operations will be conducted from the front of the building. If the fire is located in one of the residential dwellings, fireground operations will likely be conducted from the rear of the building. The entrances to the apartments are located in the rear "C" side of this building. It will be important to ensure safe fireground management and that all members operating know how the stories have been designated and which side of the building is the "A" side or front of the building.

Taxpayers/Strip Malls and Stores

Photo 4–4

Photo 4–5

Answer the following questions based on Photos 4–4 and 4–5.

1. What is the construction classification for this building?

2. What is the occupancy of this building?

3. What unique building feature is visible, in the view of the rear of the building and what size-up point will this impact?

4. What impact will the two separate 2nd floors have on our fireground operations in this building?

5. What considerations will you make for the size-up point **Exposures**?

6. What other size-up point will be a factor in the threat to the exposure buildings?

7. What will be the most efficient ways when you are on scene to identify the unusual floor layout on the 2nd floor?

ANSWERS

1. Ordinary construction.

2. The 1st floor is occupied by stores, the 2nd floor may be occupied as offices or apartments.

3. A separate independent 2nd floor area is located near the rear of the building. This will affect our size-up point **Area**; the building has an irregular 2nd floor. The 2nd floor area in the front of the building is only about one-third to one-half the length of the building.

4. It is important to identify this unique building feature early on in this firefighting operation. Access to the rear 2nd floor must be identified. Is the access to the 2nd floor area from the exterior rear door or through an interior stairway from the 1st floor? These unique building features will have an impact on fireground management and firefighter safety. For example, Firefighters become trapped on the 2nd floor of the building and call, "Mayday! Firefighter down on the 2nd floor!" Would you know where to begin to look for the firefighters?

5. The building is attached on both sides by similar type buildings. The building on the "B" side is higher. The threat to each building will depend on the location and extent of the fire in the building.

6. **Weather** — Wind direction and speed.

7. A report from the roof or a report from the rear of the building. The member operating on the roof will be able to provide the best information since the member on the ground in the rear will have limited sight distance from the ground. If members access the center portion of the roof, these members will have access into both areas of the 2nd floor through windows adjoining the roof area.

5 Garden Apartments and Townhouses

Chapter Questions

1) A construction concern in garden apartments of Class 5 — wood frame is:

   a) The use of lightweight concrete over wood floors.
   b) The use of hybrid materials in the floor and roof assemblies.
   c) Garages built under the structure.
   d) Pipe chases run from floor to floor for plumbing lines.

2) Choose the correct answer that details the size-up point of *Terrain* as relating to garden apartments and townhouses.

   a) The site topography may be sloped, auxiliary appliances may be nonexistent because of lax code requirements, and building may be set back 200–300 ft from the street.
   b) The site topography may be sloped, auxiliary appliances may be nonexistent because of lax code requirements, and narrow alleys between buildings and fenced areas make access difficult.
   c) The site topography may be sloped, buildings may be set back 200–300 ft from the street, and narrow alleys between buildings and fenced areas make access difficult.
   d) The site topography may be sloped, buildings may be set back 200–300 ft from the street, and hydrants will be present in long narrow driveways.

Fireground Size-Up Study Guide

3) **In garden apartments and townhouses, the most significant life hazard exists when?**

   a) Only the early morning hours.
   b) Late evening and early morning hours.
   c) Late evening and early morning hours, weekdays only.
   d) Late evening through the late morning hours every day of the week.

4) **Unit separation voids created by double-stud walls is typical in townhouse construction. Which of the following statements applies to fire extension in this area?**

   a) Fire extension is only a concern if the fire starts in this void space.
   b) Fire never extends into this space because the walls are fire-rated construction.
   c) There will only be vertical fire extension if the fire originates in this void space.
   d) The fire will spread to both adjoining dwelling units if the fire originates or extends into this void space.

5) **Area square footage concerns for the townhouse can be very different when compared to the garden apartment complex primarily from the fact that:**

   a) Townhouses always have a garage and a utility area on the ground level. Garden apartments do not have these spaces.
   b) Garden apartments are typically 1-floor level, while in townhouses there may be 2 or 3 floors of living space.
   c) Garden apartments always have a common interior stairway, and townhouses always have a door leading directly to the exterior.
   d) Both **b** and **c** are correct answers.

6) **The term terrace apartment refers to an apartment that is:**

   a) Partly below grade in the front of the building and open to grade at the rear of the building.
   b) A loft-type dwelling unit in the dormered area of the roof opening onto the roof.
   c) An apartment opening at grade level and does not connect to the public or common stairway of the building.
   d) A basement apartment that is more than 50% above grade with a separate outside stairway for entry to the dwelling unit.

7) **When fighting a fire in a garden apartment or townhouse, the risk to firefighter safety is greatly increased when:**

   a) The fire originates in the basement area of a garden apartment.
   b) Fire occurs in the late evening to early morning hours.
   c) The fire extends beyond the contents of the structure and begins to attack the structural members.
   d) When the fire extends from the apartment at fire origin to the public or common stairway.

8) **When occupants cannot exit via the interior stairway, the windows or balconies of garden apartments and townhouses will become an area of refuge for the occupants as they wait for rescue/removal from these areas. When the fire officer is confronted with numerous victims at these locations, removal efforts should be prioritized based on the following:**

   a) Number of ladder companies on scene or responding to the incident.
   b) The number of occupants who can be removed by aerial ladder or tower ladder.
   c) The number of occupants who can be removed with a single section 20-ft roof ladder.
   d) The areas of greatest danger.

9) **Which of the responses is correct for fighting a fire in townhouses where the lightweight wood truss floor assembly is involved in fire?**

   a) Firefighters should never operate under the floor but may cut the floor from above to expose the fire.
   b) Use a thermal imaging camera to identify the fire location and extent and operate from refuge area or areas supported around or within the building.
   c) Cut the exterior wall away to expose the fire in the concealed space and operate from the exterior of the building.
   d) The thermal imaging camera will be of little value if the fire is in the concealed floor space. Aggressive removal of the ceiling finish will reveal the location and extent of the fire.

10) **One of the primary life hazard differences between 1- and 2-family dwellings and garden apartments is:**

   a) The height of the buildings.
   b) The construction type.
   c) Fire extension probability from garages located under the building.
   d) The number of occupants in the building who may need to be rescued.

11) **Which of the following are common construction concerns for fire officers operating at a fire in garden apartments and townhouse structures?**

   a) Unit separation voids, lightweight or hybrid construction methods and materials, vertically stacked pipe chases, and back-to-back kitchens.
   b) Unit separation voids, lightweight or hybrid construction methods and materials, the height of the buildings, and back-to-back kitchens.
   c) Unit separation voids, the location of garages under the buildings, vertically stacked pipe chases, and back-to-back kitchens.
   d) Unit separation voids, lightweight or hybrid construction methods and materials, the location of firewalls or fire separation walls, and the height of the building.

12) **The most significant fire extension concern for townhouse buildings is:**

   a) Double-stud unit separation voids.
   b) Back-to-back kitchens.
   c) Vertical pipe chases running vertically from the basement to the top floor.
   d) The unenclosed interior stairway of the dwelling unit.

13) **Placing ground ladders to specific areas or targets at townhouses or garden apartment fires will be limited by:**

   a) The distance to the building from the ladder truck.
   b) The height of the building and sloped terrain.
   c) Sloped terrain and decorative landscaping.
   d) Narrow alleyways and the height of the building.

Fireground Size-Up Study Guide

Answer Key

Question #	Answer	Page #
1	b	180
2	c	192, 194
3	b	207
4	d	206
5	b	201
6	a	190
7	c	191
8	d	191
9	b	190
10	d	191
11	a	180, 183, 206
12	d	206
13	c	188

Garden Apartments and Townhouses

PHOTO SCENARIO QUESTIONS

Photo 5–1

Answer the following questions based on Photo 5–1.

1. What is the occupancy classification for this building?

2. What is the construction type of this building and what construction feature will pose a significant risk to firefighter safety?

3. Has this fire extended from the original area of fire origin, and if so, how has the fire extended?

4. What **Weather** size-up point is having an effect on this fire, and how does this relate to the **Area** size-up point?

5. What are the common street condition concerns in complexes containing residential structures such as the one in this photo?

ANSWERS

1. Multi-family residential townhouse occupancy.

2. The building is constructed of wood-frame construction; and based on the age and features of the building, it is likely that the building is constructed with lightweight wood floor and roof assemblies.

3. The fire has extended from the original area of origin; the fire began on the 2nd floor dwelling just to the left of the chimney chase with the visible fire. The fire has extended through the concealed spaces of the wood framing.

4. The wind direction and speed will have an impact on the fire, and the condition will be exaggerated if the fire extends to the attic or cockloft space of the building. As noted in the photo, the wind is blowing toward the "C" and "D" sides of this building. The building is 6 residential units wide, and each unit contains 2 dwellings arranged in an up-and-down fashion. The size of the building is significant; fortunately the wind direction will help keep the fire from extending to the larger exposed area on the "B" side since the wind is moving the fire to the "D" side of the building.

5. Common street conditions at residential developments include: limited access to the complex or development, narrow streets often clogged with cars, dead end streets, and cul-de-sacs.

Garden Apartments and Townhouses

Photo 5–2

Answer the following questions based on Photo 5–2.

1. What is the construction classification for this building?

2. What is the occupancy classification for this building?

3. How will the size-up points time and life hazard affect your firefighting priorities for fire in these type of structures?

4. How does the size-up point **Terrain** affect your fireground operations?

5. At the inside corner of the building, the roof structures change design and elevation. How can you determine if a firewall exists at this point of the building? How will this impact the size-up point area?

6. Detail three reasons why it is important to view the rear of garden apartment style buildings.

Answers

1. Ordinary construction.

2. Garden apartment type multiple dwelling.

3. The life hazard will largely be based on the time of day the incident is occurring. Fires that occur between 11 P.M. and 7 A.M. are when the most significant life hazard exists since the occupants are likely to be sleeping. Fires occurring during other hours will also represent a life hazard, but the number of occupants at risk during these hours will be reduced because most occupants who are awake will self-evacuate.

4. The terrain in the courtyard is unobstructed except for the post and rail fence that will really not impede our operations. The large setback from the street to the building at the back of the court will affect our hoseline selection and the time it will take to place a line in service.

5. Determining if a firewall exists at this location will help determine the potential fire area. A member assigned to the roof will be able to identify a firewall if the firewall extends through the roof. Another way to determine if a firewall exists will be to go to the 3rd floor apartments in the areas where the inside corner is located and open the ceilings and examine the cockloft to determine if a firewall exists.

6. The rear of the building should always be viewed for occupants, fire, and access. Occupants may be trapped at windows or on fire escapes or may have jumped and could be injured on the ground. The location of the fire may not be obvious from the front of the structure but may be very obvious if the fire is venting out the rear windows of the building. Viewing the rear of the building will also provide an opportunity to determine if firefighters can access the building via fire escape for ventilation, to conduct searches, and to determine if the fire escape will provide access to the roof. Additionally, access for fire apparatus will be determined by viewing the back of the building.

6 Row Frames and Brownstones

Chapter Questions

1) Typical row frame buildings will be _____ in height and _____ in width.

   a) 1–4 stories … 15–30 ft.
   b) 2–4 stories … 20–30 ft.
   c) 1–3 stories … 20–30 ft.
   d) 2–4 stories … 15–30 ft.

2) The practice of constructing row frame buildings with partition walls having shared joist pockets is:

   a) Of no concern to the fire officer since the wall is of masonry construction.
   b) Only likely to be a problem if fire enters the cockloft space.
   c) Likely to allow the fire to extend into the next occupancy through the shared joist pockets.
   d) Only a problem if the wood-frame walls are constructed of balloon-framing method.

3) Two construction features of row frames that will influence fire spread into adjoining exposures are:

   a) Light and air shafts and HVAC duct work.
   b) Common cockloft and balloon framing.
   c) Platform framing and common cockloft.
   d) Light and air shafts and common cockloft.

Fireground Size-Up Study Guide

4) **The size-up point Area for brownstone structures must include determining the grade level at the rear of the structure because:**

   a) Garages may be located under the buildings in the rear.
   b) A change in grade will make a 4-story building 5 stories from the rear.
   c) A separate entrance to a basement apartment will be located in the rear.
   d) Both **b** and **c** are correct answers.

5) **In row frame buildings, wind speed and direction are a concern when:**

   a) The fire is in the basement.
   b) The fire is running through the balloon-frame walls.
   c) The fire is on the 2nd or 3rd floors of the building.
   d) The fire has taken possession of the cockloft.

6) **An important factor in determining the number of stories or identifying the 1st floor of a brownstone occupancy is:**

   a) Any story from the front of the building that is partly below grade should be considered the basement.
   b) Any story from the front of the building that is more than half above street grade should be referred to as the 1st floor.
   c) Any story that is partly below grade at the front of the building but entirely at or above grade at the rear of the building shall be referred to as the 1st floor.
   d) In order for the story of the building to be considered the 1st floor, the floor must be at or above grade at the front of the building.

7) **The most significant fire spread concern affecting firefighter safety is fire spread potential via:**

   a) Common cockloft and balloon-frame walls.
   b) Common cockloft and light and air shafts.
   c) Balloon-frame walls and light and air shafts.
   d) Numerous concealed spaces and the open interior stairway within the dwelling unit.

8) *For a fire on the top floor of a row frame, the two primary means of natural ventilation are:*

   a) Skylights and light and air shaft.
   b) Skylights and roof ventilators.
   c) Skylights and boarded-over skylights.
   d) Skylights and scuttles.

9) *A quick and relatively easy method of checking for fire extension into the cockloft is to:*

   a) Poke an inspection hole in the ceiling of each room on the top floor.
   b) Remove the return walls of the scuttle opening.
   c) Cut a hole in the exterior wall of the high side of the cockloft.
   d) Cut triangular inspection holes in the roof of each exposure.

10) *Placement of the first deployed hoselines at any fire will always depend on the life hazard and the location of the fire. Barring specifics, the fire hoseline is stretched _____.*

    a) Through the front door to extinguish the fire.
    b) Through the front door to gain control of the building's interior stairway.
    c) Through the front door to the basement to confine the fire.
    d) Through the front door to the 2nd floor to assist in rescue efforts.

11) *For the size-up point **Occupancy**, the brownstone structure may also contain the following occupancy classification:*

    a) Multiple dwelling.
    b) A garage in the basement accessible from the rear of the building.
    c) A retail shop or other business on the 1st floor.
    d) The entire structure may have been converted to a doctor's office.

12) **The following considerations apply to fire apparatus placement at row frame or brownstone structures:**

   a) The street will always be a two-way street; leave enough room for the ladder company to have scrub area of the fire building and its exposures.
   b) If the street is a one-way street, leave enough room for the ladder company to reach the front of the fire building and allow enough room for the second due engine to pass the ladder company to establish a water supply.
   c) The ladder company should always enter the block first to ensure access to the front of the building, and the street may be a narrow one-way street.
   d) If the street is a one-way street, leave enough room for the ladder company to have scrub area at the fire building and its exposures.

13) **Forcible entry concerns for both row frame and brownstone structures follow the forcible entry options at most residential structures with the addition of the following:**

   a) An ornamental steel security gate may be present on the front door.
   b) Roll-down security gates may be installed on the rear windows and doors on the basement and front floor levels.
   c) Security gates may be installed over the windows at the basement or 1st floor level.
   d) Both answers **a** and **b** are correct.

Fireground *Size-Up Study Guide*

Answer Key

Question #	Answer	Page #
1	b	213
2	c	235
3	d	231
4	b	233
5	d	229
6	b	226, 227
7	d	223
8	d	221
9	b	221, 222
10	b	219
11	c	217
12	d	220
13	c	220, 221

7 Churches

Chapter Questions

1) **Hanging ceiling spaces in churches create unusual firefighting problems for which of the following reasons?**

   a) The space may be a large combustible concealed space with limited access.
   b) HVAC equipment is present, and the area is loaded with combustible dust.
   c) The space is always high from the floor, and you cannot identify these spaces from the interior of the church.
   d) Access to the hanging ceiling is through the exterior wall. This requires additional resources.

2) **Occupancy of church buildings is not simply limited to services on Saturdays, Sundays, and holidays. Other uses may include the following:**

   a) Day care centers, banquet halls.
   b) Homeless shelters.
   c) Ministers' residence.
   d) All of the above.

3) **Fire in the church proper is easier to locate and access, but the following challenges are present with a fire in this area:**

   a) Limited access for aerial apparatus and large combustible area.
   b) Large open area with combustible construction and interior finish.
   c) Ventilation of this space is a problem due to limited openings.
   d) Advancement of hoselines is difficult due to terrain and large setbacks.

55

Fireground Size-Up Study Guide

4) **The collapse potential of a church steeple may impact surrounding buildings that lay in the collapse zone of the steeple. Which of the following is consideration for the collapse of the steeple?**

   a) Keep fire apparatus and manpower away from the front of the steeple, use a transit to detect movement in the steeple.
   b) Establish collapse zones in all directions for the full height of the steeple. Evacuate structures within the collapse zone and use a transit to detect movement in the steeple.
   c) Establish collapse zones in all directions for the full height of the steeple. Use a transit to detect movement of the steeple and evacuate structures in the collapse zone if movement is detected.
   d) The steeple will usually collapse onto the roof of the church. Establish collapse zones in the direction nearest the church roof. Use a transit to detect movement of the steeple.

5) **Two fireground safety concerns while operating at the exterior of church fires include:**

   a) Steep terrain and large numbers of stairs.
   b) The use of numerous large caliber hose streams during defensive operations and icing conditions during cold weather.
   c) Large slate tiles may be dislodged and fall on firefighters and exterior walls may collapse on firefighters.
   d) The potential for the concealed space of the steeple to backdraft and opposing hose streams from ladder pipe operations.

6) **The construction class associated with gothic style churches is:**

   a) Class 1 — fire-resistive construction.
   b) Class 3 — ordinary construction.
   c) Class 4 — heavy timber.
   d) Class 5 — wood frame.

Churches

7) *Which answer is correct in describing problems with below-grade areas of churches?*

   a) Ventilation will be limited, if possible at all. Often access to the below-grade areas will be through a narrow stairway, and hoseline stretches to the below-grade areas will be lengthy and time-consuming.
   b) Ventilation will be limited, if possible at all. Often access to the below-grade areas will be through a narrow stairway, and there may be numerous avenues for the fire to spread from this area to other areas of the building.
   c) Hoseline stretches to the below-grade areas will be lengthy and time-consuming. Searches of these areas must always be delayed until the fire is under control and the below-grade area may be used as a church social hall.
   d) Hoseline stretches to the below-grade area will be lengthy and time-consuming. Often access to the below-grade areas will be through a narrow stairway and the below-grade area may extend to other structures attached to the church, creating large confusing floor areas.

8) *In addition to the concern of large open spaces in churches, firefighters must also be concerned about:*

   a) Heavily fortified entry doors and windows.
   b) Complex fire suppression systems located in these buildings.
   c) Identifying the roof structure type as this is often difficult in church buildings.
   d) Maze-like areas in and around the building where companies may have to operate.

9) *The most inaccessible spaces in a church are:*

   a) Below-grade areas.
   b) Roofs with a steep pitch and slate roof coverings.
   c) The hanging ceiling space over the church.
   d) The areas behind and above the altar.

10) *A key factor in considering exposure problems from church fires is:*

   a) Wind speed and direction.
   b) The tremendous amounts of radiant heat that can be produced.
   c) Wind speed and construction of the exposure.
   d) The tremendous amounts of convected heat that can be produced.

11) **A key factor in firefighter safety when conducting offensive interior firefighting operations at a church is:**

   a) Obtain immediate information on the fire involvement of the void spaces.
   b) Identify the interior access to the hanging ceiling.
   c) Identify the type of roof structure on the building.
   d) Obtain immediate information on the fire's location if the fire is in a below-grade space.

12) **When the decision is made to commit engine companies arriving at a church fire to an offensive attack, take these actions:**

   a) Call additional resources to help establish a sufficient water supply and prepare ladder companies for master stream operations in the event the offensive attack fails.
   b) Stretch 1¾-in. or 2-in. hoselines to the seat of the fire, since this size line will allow greater maneuverability and initiate ventilation so the hoseline can advance to the seat of the fire.
   c) Stretch large hoselines to all areas identified as possible areas of fire extension and initiate ventilation.
   d) Establish a sufficient water supply and stretch large hoselines to the seat of the fire.

13) **The primary concern for firefighters operating in churches that have vaulted ceilings with heavy ornamental plaster is:**

   a) The problem associated with exposing hidden fire behind these large cross-sections of plaster.
   b) The probability of large sections falling and dropping to the floor as fire attacks the attic space.
   c) Fire extension to the space above the plaster will go undetected because of the large cross-sections of plaster.
   d) Both answers **a** and **c** are correct.

Answer Key

Question #	Answer	Page #
1	a	244
2	d	246, 247
3	b	264, 265
4	b	266, 267
5	c	254, 255
6	c	244
7	b	264
8	d	262
9	c	265
10	b	261
11	a	253
12	d	249
13	b	245

Churches

PHOTO SCENARIO QUESTIONS

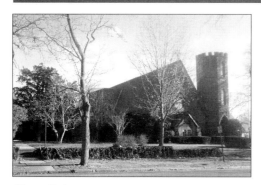
Photo 7–1

Photo 7–2

Photo 7–3

Based on Photos 7–1, 7–2, and 7–3, answer the following questions.

1. What is the most likely type of construction for this building?

2. The roof structure on this building is most likely what?

3. Based on Photo 7–2, what other occupancies or uses may exist in the building?

4. Based on Photos 7–1 and 7–2, what area and terrain concerns exist at this building?

5. Detail your size-up concerns for **Terrain.**

Answers

1. Based on the age and architecture of the building, the building will likely be a hybrid of ordinary and heavy-timber construction.

2. A truss roof assembly.

3. The wing on the "D" side of the building is 2 stories and is of substantial square footage certainly large enough to house several ancillary uses including but not limited to the following: day care center, Sunday school, church social hall, etc.

4. The terrain drops off as you move toward the intersecting street on the "D" side of the building. This reveals an additional occupied story at this lower grade. This change in grade and additional story at this level may make fireground management more difficult if the stories and levels are not properly identified. A positive factor about the below-grade story is that access to the basement or crawl area under the sanctuary can be used to attack any fire in this location from an access point from grade level.

5. Photo 7–1 depicts a deep setback from the street, but the area is free of overhead obstructions. Photo 7–2 depicts there is good access to the wing on the "D" side from the intersecting street, and there are no overhead obstructions on this street. Photo 7–3 depicts poor access to the rear of the structure and a fence. The slope of the terrain will prohibit vehicle access. The proximity of the adjoining, structure will preclude operation of hoselines from this area as the firefighters would be operating in the collapse zone.

Churches

Photo 7–4

Photo 7–5

Based on Photos 7–4 and 7–5, answer the following questions:

1. What is the occupancy of the building?

2. How does the occupancy affect the **Life Hazard** point of size-up?

3. What is the construction class of this building?

4. What type of roof structure might you expect to find on the building?

5. What concerns may exist for the size-up point **Street Conditions**?

6. What are your considerations for the **Area** size-up point?

7. If upon your arrival, the fire was located on the 1st floor of the attached dwelling, what will be your concerns for fire extension?

8. If upon your arrival, the fire was located in the utility room located at the rear, lower level of the church, what will be your concerns for fire extension?

ANSWERS

1. The building is a church with a residence attached to the "B" side.

2. The life hazard for the residential dwelling will be high at night when the occupants are sleeping and the life hazard for the church will be high during the day when services are conducted. Additionally the church hall in the basement may be used at all different times for different church and community functions.

3. Wood-frame construction.

4. Wood truss roof on the church and wood rafter roof on the residence.

5. The street is not very wide, therefore apparatus placement is important. Keep in mind the ladder company must have access to the front of the building, and water supplies from hydrants will have to be established. The overhead wires will limit the use of the aerial ladder at the front of the building. Access to the street and the parking lot on the "D" side will be further limited by the presence of additional cars present during services or other functions occurring at the church.

6. The size of the church and the attached dwelling are relatively small.

7. As in most dwelling fires, the primary avenue of fire extension is via the interior stairway. Additionally, you must quickly identify the areas or walls of the building connected to the church and assign crews to these areas to check for or hold fire extension in check.

8. Vertical fire extension through the balloon-frame structure is the primary concern for vertical fire extension. Based on the type and age of the structure, the stairway between the basement and the 1st floor will likely be open and constructed of wood. This stairway will be a likely path for fire extension. You must also quickly identify the areas or walls of the church connected to the residential dwelling and assign crews to these areas to check for or hold fire extension in check.

8 Factories, Lofts, and Warehouses

Chapter Questions

1) **Loft buildings usually house:**

   a) One tenant or operation that includes their manufacturing facility and storage warehouse under one roof.
   b) A number of different tenants or manufacturers, all within one building.
   c) Manufacturing operations on all floors except the top two floors of the building, which are occupied as residential dwelling units.
   d) Manufacturing operations and warehouse space on all floors except the top two floors of the building, which are occupied as residential dwelling units.

2) **Ventilation responsibilities will focus on:**

   a) The fire's location and extension probability.
   b) The location of any natural openings and the height of the building.
   c) The fire's location and the location of any endangered occupants.
   d) The location of any natural openings and the location of any endangered occupants.

3) **The most common firefighter safety concern while operating at factories, lofts, and warehouses is:**

   a) Disorientation.
   b) Collapse of the roof structure.
   c) Establishing and maintaining proper collapse zones.
   d) Depletion of air in the firefighter's SCBA when the firefighter has progressed deep into the structure.

Fireground Size-Up Study Guide

4) **Older factory buildings will typically be:**

   a) Class 3 — ordinary construction, 4 to 6 stories high.
   b) Class 4 — ordinary construction, 2 to 4 stories high.
   c) Class 3 — heavy timber construction, 2 to 4 stories high.
   d) Class 4 — heavy timber construction, 4 to 6 stories high.

5) **Newer warehouse buildings will have which of the following construction characteristics?**

   a) Class 1 — construction with masonry or tilt-up walls and an open steel bar joist/metal deck roof assembly.
   b) Class 2 — construction with masonry or tilt-up walls and an open steel bar joist/metal deck roof assembly.
   c) Class 4 — construction with masonry block walls and a lightweight wood truss assembly.
   d) Class 2 — construction with steel panel exterior walls and open steel bar joist/metal deck or lightweight wood truss roof assemblies.

6) **Because of the large floor areas in factories, lofts, and warehouses, the best search method employed to safely and successfully search these buildings is:**

   a) Perimeter search method.
   b) Light guided search.
   c) Rope search method.
   d) Combination perimeter search and light guided search method.

7) **Which of the following statements is true concerning the size-up Height of newer warehouse buildings:**

   a) Height is never a concern because you can always reach the roofs of these buildings via aerial ladder.
   b) Some 1-story warehouses may actually be 30–40 ft above street level, equaling the height of a 2- or 3-story building.
   c) The number of stories may change as the grade or slope of the property changes. A 1-story building from the front may be 2 stories at the rear of the building.
   d) The true height of the building or roof may not be truly identified, as these buildings typically have parapet walls extended 3–6 ft above the roofline.

8) **Open interior staircases in older factory buildings are a concern because:**

   a) Depending on the size of the building, there may only be one interior stairway that serves the entire building.
   b) The interior stairway is usually at the rear of the building, the most remote point for firefighters to access.
   c) Smoke entering the stairway will cause difficulty for anyone attempting to exit down, as well as for firefighters attempting to enter and climb up.
   d) Both answers **a** and **c** are correct.

9) **Loft and older warehouse buildings are often converted to other use or occupancies, most commonly for residential use. When presented with this conversion, your size-up considerations change most notably in the areas of:**

   a) Life hazard and apparatus and staffing.
   b) Life hazard and the location and extent of the fire.
   c) Life hazard and height of the building.
   d) Both answers **a** and **c** are correct.

10) **Which of the following structural components may be found on the interior and exterior of loft buildings?**

   a) Steel columns.
   b) Wood columns.
   c) Masonry cornices.
   d) Cast iron columns.

11) **Which response is correct when considering hoseline selection at fires in factories, lofts, and warehouses?**

   a) The mobility of the hoseline, the building's square footage, and fire extension probability.
   b) The building's square footage, life hazard potential, and hose stream reach, volume, and penetration.
   c) The building's fire load, the building's square footage, and hose stream reach, volume, and penetration.
   d) The mobility of the hoseline, location of the fire in the building, and fire extension probability.

12) **Forcible entry into factories and warehouses has been historically difficult and time-consuming for which of the following reasons?**

   a) Heavy swinging gates, roll-down steel doors, and chain link fences around the perimeter of the building.
   b) Additional truck and rescue companies are required, since special power tools may be needed to force entry into multiple openings.
   c) Heavy swinging gates, roll-down steel doors, and steel shutters.
   d) Additional staffing are required to ensure all exterior openings are forced. This must take place simultaneously with advancement of hoselines and ventilation operations.

13) **A firefighter safety concern for operating in high-ceiling warehouse occupancies is:**

   a) The danger associated with falling from a mezzanine in the building.
   b) Fire conditions above at ceiling or roof level may be drastically different from the conditions encountered at floor levels.
   c) The danger of objects such as light fixtures and unit heaters falling from the high ceiling.
   d) Water from discharging sprinkler heads may soften the bottom of pallet loads of stock, causing the unstable piles to fall on firefighters or block their path of exit.

14) **The stock arrangement and method of storage in warehouse occupancies have an impact on fire growth. Which of the following statements is true?**

   a) Stock stacked vertically enhance the rate of burning by aligning the fuel with the convection column.
   b) In warehouses with rack storage, more narrow aisles means fire will more likely spread via convected heat currents to the adjoining racks of materials.
   c) Stock stacked vertically and separated by aisles that are 8 ft or more wide will reduce the potential for horizontal fire spread.
   d) The stock arrangement and method of storage have minor impact on fire growth; the key factor in fire growth is the combustibility of the stock.

Factories, Lofts, and Warehouses

15) **Large area factories, lofts, and warehouses may contain firewalls designed to act as a barrier to fire. Which of the following statements are true regarding fire operations in buildings with firewalls?**

   a) A crew should be assigned to close all fire doors in the building as soon as staffing permits.
   b) Openings in the firewalls must be protected to prevent fire from crossing through, and self-closing fire doors must be checked to ensure they are functioning as intended.
   c) A crew should be assigned to chock open all fire doors in the building. This facilitates hoseline advancement and ventilation.
   d) Most firewalls have been negated by the numerous openings created during renovations, therefore firewalls will have no impact on firefighting strategies.

Answer Key

Question #	Answer	Page #
1	b	272
2	c	281
3	a	283
4	d	273
5	b	274, 275
6	c	282, 283
7	b	304
8	d	302
9	b	277
10	d	274
11	c	279
12	c	282
13	b	285
14	a	285
15	b	299

Factories, Lofts, and Warehouses

PHOTO SCENARIO QUESTIONS

Photo 8–1

Based on Photo 8–1, answer the following questions about this building and occupancy.

1. What is the occupancy of this building?

2. What is the construction type for the building?

3. What area concerns are revealed in the view of the building?

4. Based on your answer to question #2, what type of roof structure on this building will create a large safety concern for firefighters operating on or under this roof?

5. What **Street Condition** and **Terrain** size-up concerns are present in this view of the building?

6. What is the most common auxiliary appliance found in this type of occupancy?

7. What occupancy and special considerations might be an important part of your size-up at this building?

71

Fireground Size-Up Study Guide

ANSWERS

1. The occupancy appears to be some type of light manufacturing with an office area in the front of the building.

2. The front office area appears to be ordinary construction; the remaining larger area of the building housing the manufacturing area appears to be non-combustible type construction.

3. There appears to be a firewall separating the office into two areas, there appears to be a firewall separating the office area from the manufacturing area, and there appears to be a firewall separating the manufacturing area into two areas. It also appears from this view that two additions have been added to the building—one at the rear, which is slightly higher than the main part of the building, and a smaller addition on the "D" side of the building identified by the shed type roof.

4. Metal Bar Joist.

5. The street width is good, but overhead obstructions are present in the form of primary wires. Access to the "D" side of the building will be slightly delayed while the locked gate is forced open. A terrain concern for access to the "D" side of the building would be how much traffic is present during working hours. This area may be clogged with trailers waiting to make deliveries or be loaded with outgoing materials.

6. Automatic fire sprinkler system. This system should be identified, and supporting this system should be a priority for the responding companies.

7. Occupancy/hazard relationship may be useful in identifying any hazardous materials or other unusual hazards that may be located in the building. This will be helpful in determining if additional or special resources will be required at this incident.

Factories, Lofts, and Warehouses

Photo 8–2

Photo 8–3

Photo 8–4

Photo 8–5

Based on Photos 8–2, 8–3, 8–4, and 8–5, what size-up points would you use to determine your fireground strategy and tactics if you were to arrive at the building at 2 A.M. with heavy black smoke visible around the roofline and out of the windows?

Answers

Construction — the building is most likely ordinary construction, exterior masonry walls with a timber truss roof.

Occupancy — the building is a warehouse used for the storage of plumbing and heating supplies.

Area and **Water Supply** — the building is large in area but appears to be only 1 story high. If this building is or becomes well involved, significant volumes of water will be required to control and extinguish this fire.

Auxiliary Appliances — the building is provided throughout with an automatic sprinkler system. The fire department connection is noted by the arrow in Photo 8–3 and is detailed in Photo 8–5. Supporting the fire department connection will be a primary consideration for this incident. The location of the fire department connection requires a long hose stretch, and this connection may not be accessible at any given time due to the parking of trucks by the loading bays.

Life Hazard and **Time** — at 2 A.M., the occupant life hazard for this occupancy will be low as the building is not open for business. Our primary life hazard will be our firefighters operating in or around the building. Concerns for firefighter safety include: (1) large warehouse areas can disorient firefighters, (2) firefighters are required to operate in large confusing areas with limited amounts of breathing air, and (3) the timber truss roof will be a concern for firefighters operating on the roof, in the building, or outside the building within the collapse zone during defensive operations.

Street Conditions and **Terrain** — the street is a wide major roadway, and operation of aerial devices is limited from the street because of the utility poles and primary electric wires. Access to the "B" side of the building appears to be good, but that may change based upon the number of trucks and other vehicles in the parking lot. If companies are placed in this parking lot, they must be situated outside the collapse zones.

Special Considerations — based on the occupancy hazard relationship, you can expect to find many hazardous materials associated with the plumbing and heating industry including but not limited to: (1) flammable and combustible materials such as glues and solvents, (2) compressed gasses and compressed gas cylinders, and (3) many different plastic and synthetic materials used in the plumbing and heating trades.

9 High-Rise

Chapter Questions

1) **The term core construction with regard to high-rise buildings means what?**

   a) All the dead load of the structure is supported by the core beams and columns.
   b) All the building services, utilities, elevators and stairs are contained in a core area, the center core design being most common.
   c) Buildings constructed with precast concrete planks for the floors. All utility openings must be "core" drilled through the concrete plank.
   d) All the wind resistance and shear bracing of the building are achieved through the core beams and columns.

2) **Class 1 — fire-resistive buildings are constructed, arranged, and protected to resist fire. By the very nature of this design, the buildings will:**

   a) Contain the fire to a small area, allowing greater evacuation time for the occupants.
   b) Always permit a total burnout of the contents without sustaining any local or major structural failure.
   c) Hold and retain the heat from the fire.
   d) Permit greater opportunities for ventilation because of the large windows in the exterior walls.

Fireground Size-Up Study Guide

3) **When elevators are in the firefighter's service mode of operation, the elevators:**

   a) Must never be left unattended. A firefighter must always stay in the elevator car to operate the car.
   b) Can still be called to the fire floor by firefighters who operate the fire service key from the elevator lobby on the fire floor.
   c) Can be called to any floor under the fire floor by any occupant of the building to assist in evacuation.
   d) May be subject to malfunction or shut down from water runoff, heat, moisture or electric failure.

4) **Which of the following statements is true regarding firefighter safety and search practices in high-rise buildings?**

   a) Search teams with a minimum of 4 firefighters should perform light scan searches in all open floor areas. Teams of 2 firefighters should perform perimeter searches in confined occupancies such as residential occupancies.
   b) In large open areas, rope search techniques should be employed. In confined occupancies such as residential occupancies, patterned searches will play a vital role; all searches should include the use of a thermal imaging camera.
   c) All searches in high-rise buildings should be conducted employing rope search techniques. This will ensure the greatest degree of firefighter safety.
   d) Large floor areas should be broken down into smaller geographical areas and should be searched using a combination perimeter and light scan search techniques using a thermal imaging camera.

5) **The use of elevators in high-rise buildings when the fire is on an upper floor will:**

   a) Reduce the lead or reflex time it takes to go to work at a fire well above the street.
   b) Aid in the evacuation of residents from the upper floors of the building.
   c) Help ensure firefighters are physically capable of mounting an initial and sustained attack on the fire.
   d) Both answers **a** and **c** are correct.

6) **Understanding the operation of a building's HVAC system is critical to the management of a fire in a high-rise building. This will help the responding fire department:**

   a) Shut down the HVAC system. This is always the first step in controlling smoke in the building.
   b) Prevent the HVAC from intensifying the fire, limit smoke spread, vent smoke, or purge unaffected areas.
   c) Identify the fire floor (after shutting the system down) and purge all floors above the fire using the HVAC. This will enhance the occupants' life safety on the floor above the fire.
   d) Shut down the HVAC system. The fire department will be able to use the system when a qualified building engineer or fire safety director responds to operate the HVAC system.

7) **Fire extension concerns from curtain wall design and poke-through construction are a greater potential in which type of occupancy?**

   a) Residential apartment high-rise buildings.
   b) Office high-rise buildings.
   c) Hotel high-rise buildings.
   d) Both answers **a** and **c** are correct.

8) **As it relates to fire spread in high-rise buildings, which of the following high-rise occupancies will provide the greatest containment of a fire?**

   a) High-rise residential apartments.
   b) High-rise office occupancies.
   c) Multi-floor townhouse style apartments in a high-rise building.
   d) Residential hotel units in a high-rise building.

9) **When it comes to the question of stairs within a high-rise building, there are a number of concerns that need to be answered. Which of the following do you need to verify?**

   a) How many stairways are in the building, what is the fire rating of each stairway, and do all stairways serve every floor of the building.
   b) What is the rating of each stairway, do all stairways serve every floor of the building, and is there access to the roof from any stairway(s).
   c) Do all stairways serve every floor, is there access to the roof from any of the stairways, and how many stairways are in the building.
   d) Is there access to the roof from any of the stairways, how many stairways are in the building, and are there any exterior stairways on the building.

10) **Pressure-reducing valves or devices are often found on the standpipe systems in high-rise buildings. These valves and devices are a concern to the fire service because:**

    a) The engine crew must set or adjust the pressure on the valve prior to operating the valve.
    b) The valve or device may not be set for the required discharge pressure. This will produce insufficient flows to the fire department hoselines.
    c) If the valve or device is hampering adequate discharge pressure, the engine crew must have the tools available and can adjust the valve or device during firefighting operations.
    d) The engine crew should always remove or bypass these devices prior to connecting the hoseline to the standpipe outlet.

11) **A proactive approach to fighting a fire in a high-rise building will be to:**

    a) Conduct extensive preplanning of all high-rises in your district before preplanning any other occupancy.
    b) Ensure all the building occupants are aware of evacuation procedures. This can be achieved through semi-annual fire drills at each high-rise building.
    c) Conduct extensive live training evacuations at each high-rise building and understand the capability and limitations of the building's fire protection systems.
    d) Provide an increased level of firefighters responding into the incident at the receipt of the alarm.

High-Rise

12) Older "heavyweight" high-rise buildings were built during the period prior to _____?

   a) 1940
   b) 1943
   c) 1945
   d) 1947

13) The most problematic building system in high-rise buildings for firefighters responding to modern high-rise buildings is the:

   a) Automatic sprinkler and standpipe system.
   b) Building HVAC system.
   c) Building elevators.
   d) Building fire pump.

14) Firefighter safety concern(s) in high-rise buildings when the fire involves the plenum is:

   a) The metal wire and grid system that holds the ceiling in place may fail, dropping the ceiling tile, grid work, lighting fixtures, phone, and cable wires onto firefighter(s) below.
   b) The fire may extend to the floor above through openings that were not firestopped and through the opening created by the curtain wall.
   c) The fire may extend behind the crew, cutting off their means of escape.
   d) Both answers a and c are correct.

15) If windows must be opened or broken during a fire in a high-rise building, this must be accomplished under the direct control of the incident commander and the operations officer because this will:

   a) Enable the occupants above the fire to remain in place if the windows on the fire floor are removed or opened.
   b) Make fire attack easier if the windows are opened on the downwind side of the building, keeping fresh air at the back of the advancing hose crew.
   c) Rain down large, heavy, lethal pieces of glass and may effect fire and smoke conditions in the building.
   d) Windows on the upper floors of high-rise buildings should never be opened or broken as you can never predict the wind speed or direction at the upper stories of the building.

Fireground Size-Up Study Guide

ANSWER KEY

Question #	Answer	Page #
1	b	314, 315
2	c	324
3	d	324
4	b	324, 325
5	d	346
6	b	344
7	b	343
8	d	342
9	c	347
10	b	332, 333
11	d	318
12	c	310
13	b	315
14	d	325
15	c	326

High-Rise

PHOTO SCENARIO QUESTIONS

Photo 9-1

Photo 9-2

Photo 9-3

Based on Photos 9–1, 9–2, and 9–3 of this high-rise building, answer the following questions.

1. What construction type is the building?

2. What type of exterior wall is present on this building?

3. What are the two most likely paths for vertical fire extension created by the construction of the building?

4. What is the primary occupancy of this building and what other occupancies need to be considered in our life hazard size-up point?

5. How does the use and age of this building affect the size-up points of area and extent of fire?

6. What terrain concerns exist based on the views of this building?

7. What building system is likely to be the most problematic for the responding fire companies and is likely to fail early in the incident?

8. List the auxiliary appliances likely to be found in a high-rise building.

Fireground Size-Up Study Guide

ANSWERS

1. Fire-resistive.

2. Curtain wall.

3. Poke-through assemblies in the floors and voids created between the floor slab and the curtain wall.

4. The primary occupancy of this building is an office building. Photo 9–2 depicts a day care center located on the 1st floor of the building.

5. The building is a modern high-rise office building. These occupancies usually have large open floor areas using low office partitions to create workstations. This type of arrangement creates large fire areas without smaller rooms that would tend to limit fire spread.

6. Access to the front of the building appears to be adequate; but during business hours, it can be busy with cars and delivery vehicles. Access for aerial devices will be limited at the rear of the structure because of the parking deck located at the rear of the building.

7. The building elevators.

8. They include:

   - Automatic sprinkler system.
   - Building fire pump.
   - Standpipe system, where outlets may contain pressure-regulating devices.
   - Automatic fire alarm system.
   - Voice evacuation system with a public address system that can be used by the fire department.
   - An internal fire department communication or phone system.
   - A smoke control system that may or may not be part of the building HVAC system.

10 Vacant Buildings

Chapter Questions

1) **Which of the following statements is true of vacant buildings?**

   a) A vacant building that has sustained minor damage from one previous fire is of no concern to the fire department.
   b) A previous fire of any size could have done considerable damage to a building; the structural integrity is in question.
   c) The fire department only needs to be concerned when a vacant building has suffered numerous fires, because multiple fires may affect the building's structural integrity.
   d) Firefighting strategies in vacant structures will only change if previous fires have burned away major portions of the building's roof.

2) **Which of the following statements applies to fires in multi-story vacant buildings?**

   a) Fire extension potential in vacant buildings will be the same as for occupied buildings.
   b) The fire extension potential has changed because of the removal of piping, wiring, and plumbing fixtures.
   c) Vandalism, weather damage, and previous fires will add to fire extension probability in vacant buildings.
   d) Both **b** and **c** are correct.

Fireground Size-Up Study Guide

3) **Vacant buildings where the openings have been secured or fortified present the following problems to responding fire companies:**

   a) Delayed discovery of the fire, increased risk from flashover, and great extension probability as these fires always start in the basement.
   b) Firefighters may be easily disoriented from the lack of exterior openings.
   c) Fire companies will be delayed in effective extinguishment of the fire, and the building may lack openings for secondary egress.
   d) Both **b** and **c** are correct.

4) **Firefighters should avoid using fire escapes on vacant buildings for the following reasons:**

   a) Years of neglect and disrepair may cause the steel to corrode and rot, which could affect the integrity of the fire escape.
   b) The windows accessing the fire escape are always blocked up so there is no reason to use the fire escape.
   c) The best way to access the upper floors in a vacant building is via the interior stairway. This stairway is always structurally sound because it is in the center of the building away from openings.
   d) Both **a** and **b** are correct answers.

5) **A vacant building marking system should contain the following information:**

   a) Structural damage, owner phone number, and any missing stairs/landings.
   b) Structural damage, condition of fire escapes, and condition of automatic sprinkler/standpipe system.
   c) Structural damage, roof openings, and holes in floors.
   d) Structural damage, roof openings, and number of previous fires.

6) **The easiest method of updating the marking system for vacant buildings is:**

   a) Notify the fire prevention division to update the building's condition.
   b) Update the marking system prior to leaving the scene of a fire in a vacant building.
   c) Notify the next shift to update the marking system when the exterior walls are dry and can be painted.
   d) The city building department should determine the structural integrity of the building and make the proper changes to the marking system.

7) **Which of the following statements are most true about fires in vacant structures?**

   a) Vacant structures are never occupied, so the only **life hazard** concern is that of the firefighters.
   b) Fire officers and firefighters need to take a well-calculated and cautious approach and most often should conduct defensive operations only.
   c) If the vacant building has never been the site of a previous fire, the fire officers should employ a strategy as if the building was occupied.
   d) Fires in vacant structures require fire officers and firefighters to take a well-calculated and cautious approach before going to work in these types of buildings.

8) **Openings in previously sealed vacant buildings may be an indication that:**

   a) The building is undergoing renovations.
   b) The building may be re-inhabited or occupied by squatters or the homeless.
   c) The building is being stripped of plumbing pipe for salvage.
   d) All of the above.

9) **As an example of occupancy/construction association, if the vacant building was originally occupied as a supermarket, you may expect the features of the structure to include:**

   a) Open conveyer shafts connecting the basement to the 1st floor.
   b) A timber truss roof assembly.
   c) Refrigeration and freezer units and their associated equipment located throughout the building.
   d) All of the above are correct.

10) **Rapid fire spread in vacant buildings must be anticipated because:**

   a) All exterior openings have been sealed up and delay transmitting the alarm to the fire department.
   b) All exterior openings have been sealed up, and previous fires have weakened the structure.
   c) Numerous openings are created in the interior of the building, and the lack of exterior openings due to sealing up of these openings will cause a delay in notification to the fire department.
   d) Numerous openings created in the interior of the building, and deterioration of the floors from the elements entering through holes in the roof will lead to rapid fire spread through the building.

11) **One of the key factors in firefighter safety for fires in vacant buildings is for the first arriving officer to consider:**

   a) The original occupancy of the building.
   b) Collapse zones possibilities.
   c) Construction type of the structure.
   d) The attack mode to be used to fight the fire.

12) **Exposures become a real concern when confronted with a fire in a vacant building because:**

   a) The exposure will be vacant.
   b) Of the possibility of an advanced fire condition upon arrival.
   c) This is only a concern if the exposure is of Class 5 — wood-frame construction.
   d) Additional resources will be required if the exposure building is threatened or involved.

13) **For operations at larger vacant buildings, the incident commander should get a report from the rear of the building to help identify:**

   a) Irregular-shaped buildings, square footage of the building involved, and exposure problems not identifiable from the street.
   b) Irregular-shaped buildings, square footage of the building involved, and access to the roof via fire escapes.
   c) Irregular-shaped buildings, construction type, and exposures not identifiable from the street.
   d) Irregular-shaped buildings, access for fire apparatus, and access to the roof via fire escape.

Fireground Size-Up Study Guide

Answer Key

Question #	Answer	Page #
1	b	355
2	d	364
3	d	369
4	a	378
5	c	358
6	b	359
7	d	355
8	b	361
9	d	362
10	d	368
11	d	366
12	b	373
13	a	374

Vacant Buildings

PHOTO SCENARIO QUESTIONS

Photo 10–1

COAL TWAS WEALTHS

Vacant buildings pose unique hazards and challenges to responding firefighters. Based on Photo 10–1, what size-up considerations would be part of your initial size-up if you were the first arriving chief officer for a fire in this building?

Answers

Location and **Extent of Fire** — where is the fire in the building? Will outside streams reach the fire for a quick knockdown? Is the fire rapidly extending?

Exposures — both exposure buildings are extremely close and both buildings are wood frame with combustible siding and windows in the walls. You will need to determine if the exposure buildings are occupied. Is there an exposure problem at the rear of the building?

Life Hazard — the basement and 1st floor openings on the front of the building are secured, but a quick survey must be made of the openings on all sides of the building as unsecured openings near ground level may be an indication that people have been in the building. A vacant building marking system is visible on the front door opening to warn firefighters of hazards associated with this building.

Street Conditions — is the street wide enough to allow more than one ladder company in proximity of the building? Can the engine companies pass other apparatus to drop supply lines?

Terrain — how will firefighters access the rear yard? Hoselines may need to be placed in service in the rear of the fire building, but how will access be made to the rear yard—through adjoining yards or through the yards on the street behind the fire building?

Water Supply — an adequate water supply must be quickly established to control this fire. Working hydrants must be located and supply lines established.

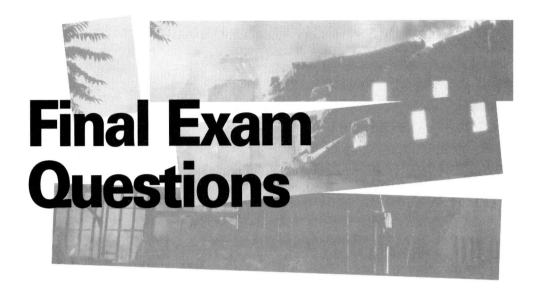

Final Exam Questions

1) **Alterations and renovations to buildings have resulted in hybrid construction. It is important for the fire officer to recognize hybrid construction because:**

   a) Hybrid construction always uses steel to replace wood structural elements.
   b) Hybrid construction may use lightweight structural elements, thereby reducing the collapse resistiveness of certain buildings.
   c) Hybrid construction is easy to recognize, so preplanning of these structures is not important.
   d) Hybrid construction uses "engineered structural elements" that have the same characteristics as the original structural components they are replacing.

2) **Steel deck roofs are usually found on buildings of:**

   a) Class 5 — wood-frame construction.
   b) Class 3 — ordinary construction.
   c) Class 2 — noncombustible/limited combustible construction.
   d) Class 1 — fire-resistive construction.

Fireground Size-Up Study Guide

3) The floors of many taxpayer-type buildings are constructed of terrazzo, marble, or poured concrete. The fire services' main concern with this type of floor is:

   a) The floor allows water to pool on the floor.
   b) These materials are often disguised when covered with tile or carpet.
   c) There may be little indication of the fire below, and a weakened floor assembly may go unnoticed.
   d) These materials are very difficult to cut in order to open the floor for ventilation purposes.

4) Firewalls in garden apartment and townhouse structures can be identified from the exterior by:

   a) Examination of the attic or cockloft space.
   b) Reviewing the building pre-plan.
   c) Firewalls cannot be identified from the exterior of the building.
   d) The extension of the firewall through the roof at the front and rear walls of the building.

5) A major exposure consideration for firefighters in townhouse buildings is:

   a) Horizontal fire extensions through pipe chases.
   b) Vertical fire extension through bathroom cabinet soffits.
   c) Horizontal fire extension through the building roof space.
   d) Horizontal fire extension through openings created for HVAC duct work.

6) Hanging ceiling spaces in churches create unusual firefighting problems for which of the following reasons?

   a) The space may be a large combustible concealed space with limited access.
   b) HVAC equipment is present, and the area is loaded with combustible dust.
   c) The space is always high from the floor, and you cannot identify these spaces from the interior of the church.
   d) Access to the hanging ceiling is through the exterior wall. This requires additional resources.

Final Exam Questions

7) Queen Anne/Victorian Style structures are examples of wood-frame construction. These structures include some of the following features:

a) They are 2½ to 3½ stories high, typically balloon-frame construction with a hip roof.
b) They are 1½ to 2½ stories high, typically balloon-frame construction with a roof containing many peaks, valleys, and dormers.
c) They are 1½ to 2½ stories high, typically balloon-frame construction with large eaves/ overhangs and a cupola on the roof.
d) They are 2½ to 3½ stories high, typically balloon-frame construction with a roof containing many peaks, valleys, and dormers.

8) If firefighters suspect the fire is overhead running the plenum space in a high-rise building, firefighters should:

a) Report this to the backup team stretching the backup line.
b) Direct a hose stream overhead periodically into this area to determine if fire is overhead.
c) Remove ceiling panels periodically ahead of their advance to determine if fire is overhead.
d) Report this to the incident commander/operations officer and retreat to the stairway.

9) When operating at a fire in a warehouse or factory, which of the following are reasons to augment the building's automatic sprinkler system?

a) It is your department's SOP. Augmenting the automatic sprinkler system helps you avoid potential litigation.
b) Many times automatic sprinkler control valves are shut down. Augmenting the building's automatic sprinkler system through the fire department connection bypasses the closed valves.
c) In older buildings, the sprinkler pipe sizes are usually too small to deliver adequate water supplies to the open sprinkler heads.
d) The sprinkler system may become taxed as the result of fire loading and stock arrangements, and the sprinkler system is only designed to deliver adequate flows to a percentage of the sprinkler heads operating.

Fireground Size-Up Study Guide

10) Which of the following statements is true of vacant buildings?

   a) A vacant building that has sustained minor damage from one previous fire is of no concern to the fire department.
   b) A previous fire of any size could have done considerable damage to a building; the structural integrity is in question.
   c) The fire department only needs to be concerned when a vacant building has suffered numerous fires, because multiple fires may affect the building's structural integrity.
   d) Firefighting strategies in vacant structures will only change if previous fires have burned away major portions of the building's roof.

11) The occupant life hazard profile for fires in multiple dwellings must follow the assigning of priority areas based on the areas of greatest danger within the building. Those areas are:

   a) Fire floor, areas connected on the same level of the fire floor, and the floor above the fire.
   b) Fire floor, floor above the fire, and floor below the fire.
   c) Fire floor, floor above the fire, and the next floor above the fire.
   d) Fire floor, floor above the fire, and the top floor of the building.

12) Row frame structures are typically Class 5 construction, using the following framing methods:

   a) Post and beam and platform-frame methods.
   b) Balloon-frame and platform-frame methods.
   c) Balloon-frame and braced-frame methods.
   d) Post and beam and balloon-frame methods.

13) Factors influencing hoseline selection and stretch for a fire in a low-rise housing project include the following:

   a) Fire floor and extent of the fire.
   b) The location of the building entrance and obstructions between the street and the building entrance.
   c) Required fire flow and staffing.
   d) Extension probability and time of day.

Final Exam Questions

14) *In wood-frame private dwellings, hazards of hybrid construction include:*

  a) The use of lightweight floor and roof trusses and unprotected structural steel components.
  b) The failure of gusset plate connectors and the use of recycled structural components.
  c) The use of unprotected structural steel to support concrete floors and poured over wood joists.
  d) The use of lightweight roof trusses with flake board on roof assemblies.

15) *Below-grade fires in taxpayers and stores are difficult and dangerous for which of the following reasons?*

  a) Low ceiling heights, lack of sprinkler systems, and poor accessibility to these areas.
  b) Limited access, heavy fire loading, and maze-like conditions.
  c) Low ceiling heights, ventilation is impossible, and heavy fire loading.
  d) Limited access, fire extension to the 1st floor is likely, and ventilation is difficult.

16) *The most common firefighter safety concern while operating at factories, lofts, and warehouses is:*

  a) Disorientation.
  b) Collapse of the roof structure.
  c) Establishing and maintaining proper collapse zones.
  d) Depletion of air in the firefighter's SCBA when the firefighter has progressed deep into the structure.

17) *The **Time of Year** size-up point is important for the following reason:*

  a) During winter months, a vacant building is more likely to be occupied by homeless person(s) seeking shelter from the winter elements.
  b) Fireground operations are more difficult during the winter months as a result of the water from master streams freezing.
  c) Fires during the summer months will be more dangerous since firefighters become more fatigued from the heat and humidity.
  d) The size-up point **Time of Year** need not be a consideration for fires in vacant buildings.

18) **The practice of constructing row frame buildings with party walls having shared joist pockets is:**

   a) Of no concern to the fire officer since the wall is of masonry construction.
   b) Only likely to be a problem if fire enters the cockloft space.
   c) Likely to allow the fire to extend into the next occupancy through the shared joist pockets.
   d) Only a problem if the wood-frame walls are constructed of balloon-framing method.

19) **Ventilation of ornamental stained glass windows may be hampered by the fact that:**

   a) The windows are expensive and irreplaceable.
   b) The windows will not be readily accessible by ground ladder.
   c) Protective mesh or grating may be installed on the exterior of the window to protect the window.
   d) Fire officers are hesitant to order these windows vented because of their historical significance.

20) **The cooling and ultimate layering of the smoke in a high-rise building is referred to as:**

   a) Stratification.
   b) Positive stack effect.
   c) Reverse stack effect.
   d) Unification.

21) **For fires in Class 5 — wood-fame structures, if the fire goes beyond the contents and begins to attack the structure, the fire officer must always consider:**

   a) Exterior walls in platform frame never have any openings between floors.
   b) In balloon-frame structures, fire extension will only be horizontal.
   c) The structural elements of Class 5 buildings are combustible, therefore the building's integrity is at risk.
   d) Additional firefighting resources will be required to locate and expose hidden fire.

Final Exam Questions

22) *Fires occurring in the 1st floor rear storage area in strip malls and taxpayers often result in:*

a) Delayed alarm and advanced fire conditions on arrival.
b) Accessibility problems since the rear door is well secured.
c) Delayed alarm and fire extension to the floors above.
d) Fire extension into the drop ceiling in the customer area of the occupancy and the cockloft space.

23) *A significant exposure concern for fires in row frames is the rear of the building. This concern in based upon the following:*

a) Rear yards separating the buildings may be a small as 20–25 ft.
b) Access to the rear yards is never a problem, so this exposure is not a concern for fire officers.
c) The exterior siding will always be combustible.
d) There is never access for fire apparatus.

24) *Understanding the operation of a building's HVAC system is critical to the management of a fire in a high-rise building. This will help the responding fire department:*

a) Shut down the HVAC system. This is always the first step in controlling smoke in the building.
b) Prevent the HVAC from intensifying the fire, limit smoke spread, vent smoke, or purge unaffected areas.
c) Identify the fire floor (after shutting the system down) and purge all floors above the fire using the HVAC. This will enhance the occupants' life safety on the floor above the fire.
d) Shut down the HVAC system. The fire department will be able to use the system when a qualified building engineer or fire safety director responds to operate the HVAC system.

Fireground Size-Up Study Guide

25) Which of the following exterior siding materials can be responsible for rapid extension of fire to exposures, upper floors of the structure, into the overhang or cornices, and into attic and cockloft spaces:

a) Asbestos siding.
b) Wood clapboard siding.
c) Asphalt siding.
d) Aluminum siding.

26) Unit separation voids created by double-stud walls is typical in townhouse construction. Which of the following statements applies to fire extension in this area?

a) Fire extension is only a concern if the fire starts in this void space.
b) Fire never extends into this space because the walls applied are fire-rated construction.
c) There will only be vertical fire extension if the fire originates in this void space.
d) The fire will spread to both adjoining dwelling units if the fire originates or extends into this void space.

27) Warehouses and factories constructed of Class 2 — limited combustible construction — will warrant serious consideration when firefighters are assigned to the roof because:

a) Roofs associated with this class of construction will be made up of combustible materials over a metal deck.
b) Roofs associated with this class of construction will be made up of lightweight concrete planks over steel "I" beams.
c) There are usually no natural openings to accomplish a quick vent opening.
d) Fiberglass panels are built into the roof decking to allow light into these occupancies. These fiberglass panels create a serious fall hazard for firefighters.

Final Exam Questions

28) *A water supply consideration for fires in vacant buildings is:*

   a) The spacing of hydrants.
   b) Water utility companies often remove fire hydrants from service in areas where vacant buildings accumulate.
   c) In areas where vacant buildings accumulate, fire hydrants are often vandalized and tampered with.
   d) Water supply considerations for fires in vacant buildings are no different than for occupied buildings.

29) *What must be an automatic thought with a fire in a tenement or non-fireproof multiple dwelling is:*

   a) Ventilation of the fire room or apartment.
   b) Positive pressure ventilation behind the advancing engine company.
   c) Ventilation of the cockloft.
   d) Ventilation of the interior stairs.

30) *Fire in the church proper is easier to locate and access, but the following challenges are present with a fire in this area:*

   a) Limited access for aerial apparatus and large combustible area.
   b) Large open area with combustible construction and interior finish.
   c) Ventilation of this space is a problem due to limited openings.
   d) Advancement of hoselines is difficult due to terrain and large setbacks.

31) *Firefighters should avoid using fire escapes on vacant buildings for the following reasons:*

   a) Years of neglect and disrepair may cause the steel to corrode and rot, which could affect the integrity of the fire escape.
   b) The windows accessing the fire escape will always be blocked up so there is no reason to use the fire escape.
   c) The best way to access the upper floors in a vacant building is via the interior stairway. This stairway is always structurally sound because it is in the center of the building away from openings.
   d) Both **a** and **b** are correct answers.

Fireground Size-Up Study Guide

32) Two fireground safety concerns while operating at the exterior of church fires include:

a) Steep terrain and large numbers of stairs.
b) The use of numerous large caliber hose streams during defensive operations and icing conditions during cold weather.
c) Large slate tiles may be dislodged and fall on firefighters and exterior walls may collapse on firefighters.
d) The potential for the concealed space of the steeple to backdraft and opposing hose streams from ladder pipe operations.

33) An important safety consideration when searching the apartment above the fire is:

a) Before entering the apartment, force another door on this floor, preferably the door directly across the hallway from the apartment to provide an area of refuge.
b) Always search the apartment immediately above the fire with a charged hoseline after the apartment has been vented. This will enable the crew to ensure the area stays tenable.
c) One firefighter should have a personal safety rope and the 2nd firefighter should have a lifeline. This equipment will ensure the crew can find their way back to the public hallway or bail out a window if conditions deteriorate.
d) The ladder company should ventilate windows of the apartment directly above the fire and leave a ladder at a window as a 2nd means of escape for the search crew. A hoseline should also be stretched to the top of the ladder to ensure the room or apartment stays tenable.

34) The term *area* is defined as the square footage involved in or threatened by the fire. The following building features should be identified to determine the area of a building:

a) Basement level area and structural support system.
b) 1st-floor level area and access to the rear of the building.
c) Irregular-shaped buildings and interconnected buildings.
d) Mezzanine areas and access to the mezzanine level.

Final Exam Questions

35) Automatic sprinkler systems are often installed in buildings housing strip malls and stores. Which of the following conditions can be detrimental to the sprinkler system's ability to operate properly?

a) Combustible void spaces, wood-frame construction, and marble floors.
b) Wood-frame construction, combustible void spaces, and inadequate system inspection and maintenance.
c) High-piled stock, access to the fire department Siamese connection, and wood-frame construction.
d) Combustible void spaces, high-piled stock, and inadequate system inspection and maintenance.

36) An important factor in determining the number of stories or identifying the 1st floor of a brownstone occupancy is:

a) Any story from the front of the building that is partly below grade should be considered the basement.
b) Any story from the front of the building that is more than half above street grade should be referred to as the 1st floor.
c) Any story that is partly below grade at the front of the building but entirely at or above grade at the rear of the building shall be referred to as the 1st floor.
d) In order for the story of the building to be considered the 1st floor, the floor must be at or above grade at the front of the building.

37) A way of preventing auto exposure in high-rise buildings to the floor above the fire is to:

a) Wash the window glass of the exposed area with an exterior master stream.
b) Remove combustibles from the exposed area and stretching and operating a hoseline on the floor above.
c) Pop open sprinkler head(s) along the exterior wall of the exposed area and operating a hoseline on the floor above.
d) Remove combustibles from the exposed area and pop open sprinkler head(s) along the exterior wall of the exposed area.

38) **The most important size-up factor in considering occupants life hazard is:**

   a) Time of the year.
   b) Day of the week.
   c) Time of the day.
   d) Occupancy classification.

39) **Area square footage concerns for the townhouse can be very different when compared to the garden apartment complex primarily from the fact that:**

   a) Townhouses always have a garage and a utility area on the ground level. Garden apartments do not have these spaces.
   b) Garden apartments are typically 1-floor level, while in townhouses there may be 2 or 3 floors of living space.
   c) Garden apartments always have a common interior stairway, and townhouses always have a door leading directly to the exterior.
   d) Both **b** and **c** are correct answers.

40) **Which of the following statements is true regarding fire attack at factories, lofts, and warehouses?**

   a) If fire is visible out of more than 3 windows on any floor of the building, a defensive fire suppression mode should be considered.
   b) If fire is visible out of more than 3 windows on any floor of the building, a blitz attack should be considered as this will provide a quick knockdown and give the occupants on upper floors of the building time to exit.
   c) Two 1½- to 2-in. hoselines is a requirement for an offensive attack. The use of two hoselines will allow for easy and quick deployment of the line through maze-like conditions while delivering sufficient water to control the fire.
   d) Hoselines capable of reaching 50–75 ft and delivering 250 gpm are required for an offensive attack.

Final Exam Questions

41) A vacant building marking system should contain the following information:

   a) Structural damage, owner phone number, and any missing stairs/landings.
   b) Structural damage, condition of fire escapes, and condition of automatic sprinkler/standpipe system.
   c) Structural damage, roof openings, and holes in floors.
   d) Structural damage, roof openings, and number of previous fires.

42) The **Life Hazard** size-up is without question the most important size-up point to a fire officer. Which of the following size-up points influence the decision-making process in determining the life hazard size-up?

   a) Construction, occupancy, and street conditions.
   b) Occupancy, height, and time of day.
   c) Street conditions, occupancy, and time of day.
   d) Occupancy, location and extent of fire, and time of day.

43) Which of the following is the best option to establish a water supply to the upper floors of a multiple dwelling if the fire department s iamese connection is clogged with debris?

   a) Run a hoseline up and over a ground or aerial ladder to the fire floor.
   b) Remove the debris. You will have time to do this as the engine crew makes its way to the fire floor.
   c) Always look for another fire department Siamese connection.
   d) Connect a supply line to the standpipe outlet on the 1st floor or other lower floor and supply the standpipe through this outlet.

44) Loft and older warehouse buildings are often converted to other use or occupancies, most commonly for residential use. When presented with this conversion, your size-up considerations change most notably in the areas of:

   a) Life hazard and apparatus and staffing.
   b) Life hazard and the location and extent of the fire.
   c) Life hazard and height of the building.
   d) Both answers **a** and **c** are correct.

Fireground Size-Up Study Guide

45) *Typical roof structure construction features of gothic-style churches are:*

   a) Scissor-type truss with asphalt-shingle roof covering.
   b) Bar-joist truss with slate-tile roof covering.
   c) King post-type truss with slate-tile roof covering.
   d) Scissor-type truss with slate-tile roof covering.

46) *When operating at fires in private dwellings, an important consideration when sizing up the **height** of the structure, peaked roofs with a pitch greater than _____ will require the use of a roof ladder. This will require greater time and resources to perform this ventilation task.*

   a) 20 degrees.
   b) 25 degrees.
   c) 30 degrees.
   d) 35 degrees.

47) *Access is usually provided to the rear of strip malls for deliveries. Access to this area may be hampered by which of the following?*

   a) Narrow roadways, dumpsters, and adjoining buildings.
   b) Delivery vehicles, dumpsters, and piles of trash.
   c) Sloped terrain, dumpsters, and building set backs.
   d) Fencing around the building, sloped terrain, and narrow roadways.

48) *The most significant fire spread concern affecting firefighter safety is fire spread potential via:*

   a) Common cockloft and balloon-frame walls.
   b) Common cockloft and light and air shafts.
   c) Balloon-frame walls and light and air shafts.
   d) Numerous concealed spaces and the open interior stairway within the dwelling unit.

Final Exam Questions

49) Pressure-reducing valves or devices are often found on the standpipe systems in high-rise buildings. These valves and devices are a concern to the fire service because:

a) The engine crew must set or adjust the pressure on the valve prior to operating the valve.
b) The valve or device may not be set for the required discharge pressure. This will produce insufficient flows to the fire department hoselines.
c) If the valve or device is hampering adequate discharge pressure, the engine crew must have the tools available and can adjust the valve or device during firefighting operations.
d) The engine crew should always remove or bypass these devices prior to connecting the hoseline to the standpipe outlet.

50) During periods of inclement weather, snowfall, or flooded roadways, if you were to arrive at a garden apartment complex and discovered a significant fire, a key consideration is:

a) Ensure the 1st due company does not get stuck and block the access roadway to the garden apartment complex.
b) The occupant(s) will be less likely to self-evacuate because of the inclement weather.
c) Ventilation and ground ladder placement will be delayed.
d) Call for additional resources early on when confronted with these conditions.

51) A collapse concern for Class 2 buildings is:

a) These buildings may also be constructed with heavy-timber truss roofs.
b) Unprotected steel that is exposed to fire will expand the steel support system, possibly pushing out the exterior walls.
c) Collapse of the curtain wall panels from the exterior of the building.
d) Additional loads to the roof structure, such as HVAC equipment for computer room.

52) **Firefighters advancing a hoseline into a store or strip mall should consider the following hazards:**

   a) Construction type, high-piled stock, and maze-like arrangement.
   b) Large floor areas, high-piled stock, and maze-like arrangement.
   c) Construction type, high-piled stock, and large floor areas.
   d) Large floor areas, poor access into the rear of the structure, and maze-like arrangements.

53) **Normally the church's occupant load will generally be at its highest when?**

   a) During Saturday and Sunday services.
   b) During weekday evening sessions.
   c) During church social events.
   d) Both answers **b** and **c** are correct.

54) **Older standpipe systems in high-rise buildings were designed to deliver _____ psi at the topmost outlet while newer standpipe systems are designed to deliver _____ psi at the topmost outlet.**

   a) 65 ... 100
   b) 65 ... 90
   c) 50 ... 65
   d) 100 ... 90

55) **Openings in previously sealed vacant buildings may be an indication that:**

   a) The building is undergoing renovations.
   b) The building may be re-inhabited or occupied by squatters or the homeless.
   c) The building is being stripped of plumbing pipe for salvage.
   d) All of the above.

56) **Which answer establishes priorities for fires in basements or below-grade areas in multiple dwellings with an open interior stairway?**

   a) Get a hoseline to the seat of the fire and vent the fire area.
   b) Use an exterior stream to knock down the main body of fire and provide ventilation at the top of the stairway.
   c) Place a hoseline in service to protect the stairway and provide ventilation at the top of the stairway.
   d) Place a hoseline in service to protect the stairway and ventilate the fire area and the 1st floor.

57) **A Hollywood front on wood-frame buildings is a concern for firefighters because:**

   a) It usually hides a light and air shaft.
   b) It disguises the actual roof construction of the building.
   c) It can be a fall hazard for firefighters who do not recognize the building feature.
   d) Both **b** and **c** answers are correct.

58) **When fighting a fire in a garden apartment or townhouse, the risk to firefighter safety is greatly increased when:**

   a) The fire originates in the basement area of a garden apartment.
   b) Fire occurs in the late evening to early morning hours.
   c) The fire extends beyond the contents of the structure and begins to attack the structural members.
   d) When the fire extends from the apartment at fire origin to the public or common stairway.

59) **Which of the following statements is true regarding occupant life safety in factory, loft, and warehouse occupancies?**

   a) There can be a limited number of people within a warehouse at any given time; a factory building with its workers will number much higher, often at all hours of the day and night.
   b) Warehouse occupancies are never open to the public; based on the small number of employees in a warehouse, the occupancy life hazard is not significant.
   c) There will always be a limited number of people within a warehouse during the day; only a security guard may be present after normal business hours as warehouse occupancies never work a 2nd or 3rd shift.
   d) There is no difference in the life hazard concern between warehouse or factory occupancies.

60) **Which of the following are considerations for rear-yard accessibility for row frame and brownstone structures?**

   a) Through an adjoining structure.
   b) Up and over the roofs of adjoining buildings.
   c) Apparatus access can always be made at the end of the block on the nearest cross street.
   d) Both **a** and **b** are correct.

61) **In addition to the concern of large open spaces in churches, firefighters must also be concerned about:**

   a) Heavily fortified entry doors and windows.
   b) Complex fire suppression systems located in these buildings.
   c) Identifying the roof structure type as this is often difficult in church buildings.
   d) Maze-like areas in and around the building where companies may have to operate.

62) **Fire officers and firefighters should attempt to determine the previous occupancy of a vacant building. This information may be useful in anticipating:**

   a) What fire department resources will be required to fight the fire?
   b) Inherent building hazards and features of the building.
   c) Conditions that may lead to rapid fire growth.
   d) Both **b** and **c** are correct.

Final Exam Questions

63) Taxpayer buildings are buildings constructed of Class 3 — ordinary construction. Often which of the following construction features were also incorporated into this type of building?

   a) Parapet walls are always constructed on the front and back walls to hide building utilities located on the roof.
   b) Steel structural elements may be built in to carry structural loads over long spans.
   c) Masonry firewalls are always located between occupancies, starting in the basement, running up and through the roof.
   d) Common basements are always present in these buildings, creating a large single below-grade fire area.

64) The following are occupant life hazard concerns:

   a) Time of day and the size of the building.
   b) Street conditions for apparatus access and terrain for ground ladder access.
   c) The number of and location of the occupants.
   d) Resources necessary and type of construction.

65) The time of day plays a major role in fires occurring in multiple dwellings. The hours of midnight to 7 A.M. are the most difficult times for fires in these buildings because:

   a) It is always more difficult to fight a fire at night.
   b) The building is most populated during these hours and many of the occupants are sleeping.
   c) Arson fires usually occur during these hours.
   d) There is usually a delay in reporting fires during these hours.

66) Placement of the first deployed hoselines at any fire will always depend on the life hazard and the location of the fire. Barring specifics, the fire hoseline is stretched _____.

   a) Through the front door to extinguish the fire.
   b) Through the front door to gain control of the building's interior stairway.
   c) Through the front door to the basement to confine the fire.
   d) Through the front door to the 2nd floor to assist in rescue efforts.

67) **Which of the following statements is true regarding the presence of spreaders on the exterior wall of a factory building?**

   a) Spreaders that are redundant across the wall of the building indicate they were part of the original design of the building. These spreaders will not be affected by a fire in the building.
   b) Spreaders that are sporadic in placement have been added to the building. They are usually installed as a precaution. If more than 3 spreaders are affected by fire conditions, the structural integrity will be in question.
   c) Failure on either side of the anchoring point or directly to the spreader itself could result in collapse of a section of the building.
   d) Failure on either side of the anchoring point or directly to the spreader itself could result in a local area collapse and will not affect the integrity of the exterior wall or interior columns of the building.

68) **For fires in private dwellings, the time of year may indicate:**

   a) Fires are more likely to start in the basement during the winter months.
   b) Occupants are less likely to be alerted by smoke detectors in the summer months because of the noise level of window air conditioning units.
   c) The number of occupants of a private dwelling may increase during holiday seasons.
   d) The time of year makes no difference for fires in private dwellings.

69) **A proactive approach to fighting a fire in a high-rise building will be to:**

   a) Conduct extensive preplanning of all high-rises in your district before preplanning any other occupancy.
   b) Ensure all the building occupants are aware of evacuation procedures. This can be achieved through semi-annual fire drills at each high-rise building.
   c) Conduct extensive live training evacuations at each high-rise building and understand the capability and limitations of the building's fire protection systems.
   d) Provide an increased level of firefighters responding into the incident at the receipt of the alarm.

Final Exam Questions

70) **Which of the responses is correct for fighting a fire in townhouses where the lightweight wood truss floor assembly is involved in fire?**

   a) Firefighters should never operate under the floor but may cut the floor from above to expose the fire.
   b) Use a thermal imaging camera to identify the fire location and extent and operate from refuge area or areas supported around or within the building.
   c) Cut the exterior wall away to expose the fire in the concealed space and operate from the exterior of the building.
   d) The thermal imaging camera will be of little value if the fire is in the concealed floor space. Aggressive removal of the ceiling finish will reveal the location and extent of the fire.

71) **A common alteration and reconstruction of brownstone structures includes:**

   a) Removal of the interior stairway.
   b) Replacement of floors with lightweight concrete over the original floor joists.
   c) Replacement of floor and roof structures with lightweight structural components.
   d) Elimination of the front exterior basement door.

72) **Buildings that are separated by narrow alleyways are immediate concerns to responding fire companies. This concern increases when:**

   a) The exposure building has combustible wood or asphalt siding.
   b) Light and air shafts are present on one or both buildings.
   c) Fire is venting out of one window of a building across the alleyway in line with a window in the exposure building.
   d) All of the above are correct answers.

73) **A special consideration for high-rise multiple dwelling fires is:**

   a) Several buildings on one site may confuse the 1st arriving companies.
   b) Check the serviceability of the standpipe system and be prepared to augment or bypass the standpipe system.
   c) Cockloft fires will be a big concern because of the height of the building.
   d) Trash and compactor chute fires are common but pose no unusual problems because the smoke and products of combustion always vent out the top of the chute.

74) **As an example of occupancy/construction association, if the vacant building was originally occupied as a supermarket, you may expect the features of the structure to include:**

   a) Open conveyer shafts connecting the basement to the 1st floor.
   b) A timber truss roof assembly.
   c) Refrigeration and freezer units and their associated equipment located throughout the building.
   d) All of the above are correct.

75) **Which response is correct when considering hoseline selection at fires in factories, lofts, and warehouses?**

   a) The mobility of the hoseline, the building's square footage, and fire extension probability.
   b) The building's square footage, life hazard potential, and hose stream reach, volume, and penetration.
   c) The building's fire load, the building's square footage, and hose stream reach, volume, and penetration.
   d) The mobility of the hoseline, location of the fire in the building, and fire extension probability.

76) **Modular/prefab construction is likely to have which of the following construction features?**

   a) Lightweight structural components, wood "I" beams, and inadequate floor systems.
   b) Lightweight structural components, wood "I" beams, and thin-gauge joist hangers.
   c) Lightweight structural components, glued-floor joist/girder connections, and thin-gauge joist hangers.
   d) Lightweight structural components, wood "I" beams, and lightweight concrete floors in bathrooms and kitchens.

77) **Which type of roof design and structure typically found on townhouse structures will necessitate operating from the protection of a bucket or aerial ladder?**

   a) Flat roof with 2 x 10 rafters supported by exterior masonry walls.
   b) Peaked roof with 2 x 8 rafters supported by wood-frame exterior walls.
   c) Peaked roof with lightweight wood-truss assemblies supported by wood-frame exterior walls.
   d) All peaked roofs, regardless of the construction of the roof assembly and supporting exterior wall type.

78) **Which of the responses below are correct factors that may prevent or inhibit total efficient evacuation of high-rise office buildings?**

   a) The occupants will try to use the elevators, and they may not be familiar with alternate means of egress.
   b) The occupants may not be familiar with alternate means of egress, and many occupants chose to ignore the building fire alarm signal.
   c) The high number of occupants in the building may prevent total evacuation, and the occupants may not be familiar with alternate means of egress.
   d) Both answers **a** and **c** are correct.

79) **Which of the following conditions or considerations will directly impact firefighter safety?**

   a) Familiarity with the structure, advanced fire conditions, contents and square footage, firefighter disorientation, and collapse.
   b) Familiarity with the structure, advanced fire conditions, construction type, extension probability, and time of day.
   c) Familiarity with the structure, advanced fire conditions, time of the year, extension probability, and number of stories/height of the building.
   d) Both **a** and **c** are correct answers.

80) **Which statement is true concerning fire detection and suppression equipment in church buildings?**

   a) The absence of automatic alarms and fire suppression equipment is not usually a problem.
   b) Most church buildings have been retrofitted with complex fire detection systems to help combat large loss from arson fires in these buildings.
   c) The absence of automatic alarms and fire suppression equipment is a major contributing factor to the large fire loss in churches.
   d) Because most church fires originate in below-grade areas, installation of automatic sprinkler systems in these areas greatly reduces the potential for a large loss fire.

81) **Oftentimes individuals employed at certain sites and buildings can be of valuable assistance by providing certain technical information about a building's systems or hazards. Which of the these is least likely to be of assistance to a responding fire department?**

   a) The plant safety engineer at a chemical facility.
   b) On-site fire brigade members at an industrial complex.
   c) The building engineer at a commercial high-rise.
   d) The custodial staff at a large shopping mall.

82) **A key factor in firefighter safety when conducting offensive interior firefighting operations at a church is:**

   a) Obtain immediate information on the fire involvement of the void spaces.
   b) Identify the interior access to the hanging ceiling.
   c) Identify the type of roof structure on the building.
   d) Obtain immediate information on the fire's location if the fire is in a below-grade space.

83) **One of the key factors in firefighter safety for fires in vacant buildings is for the first arriving officer to consider:**

   a) The original occupancy of the building.
   b) Collapse zones possibilities.
   c) Construction type of the structure.
   d) The attack mode to be used to fight the fire.

Final Exam Questions

84) **When considering the "area" of fire involvement, fires in larger private dwellings with high ceilings result in _____ while smaller private dwellings with smaller rooms and lower ceilings are more likely to _____.**

   a) More concentrated area at the fire involvement ... have the fire spread to the next story or the attic of the private dwelling.
   b) More concentrated area of fire involvement ... have the fire spread beyond the room of origin.
   c) A slower developing fire with a great potential for backdraft ... have a rapid developing fire with a great potential for flashover.
   d) A slower developing fire with a great potential for flashover ... have a rapid developing fire with a great potential for backdraft.

85) **For fires in garden apartments and townhouses, most fire departments will lead off with their department's designated attack hoseline. Most often the choice hoseline will be which of the following?**

   a) A 1½- to 1¾-in. hoseline delivering 125 to 160 gpm.
   b) A 1½- to 2½-in. hoseline delivering 125 to 250 gpm.
   c) A 1¾- to 2-in. hoseline delivering 180 to 200 gpm.
   d) A 1¾- to 2½-in. hoseline delivering 160 to 250 gpm.

86) **Apparatus and staffing size-up considerations at row frame and brownstone buildings include:**

   a) Viewing and determining access to the rear of the building.
   b) Apparatus placement and staffing.
   c) Staffing and number of buildings exposed.
   d) Number of buildings exposed and time of day.

Fireground Size-Up Study Guide

87) **Which of the following statements is true concerning ladder placement at fires in taxpayers/strip malls and stores?**

   a) An aerial ladder should always be placed to the roof in the front of the building, and one ground ladder should be placed to the roof from the rear of building.
   b) An aerial ladder should always be placed to the roof from the front or rear of the building, and ground ladders should be placed to the exposures at the front of the building.
   c) Three ladders should be placed at the front—one to the fire building, one to each exposure building, and (when practical) one additional ladder placed at the rear of the building.
   d) Two ladders should be placed to the front of the building, and (when practical) two additional ladders should access the roof from the rear.

88) **Forcible entry into factories and warehouses has been historically difficult and time-consuming for which of the following reasons?**

   a) Heavy swinging gates, roll-down steel doors, and chain link fences around the perimeter of the building.
   b) Additional truck and rescue companies are required, since special power tools may be needed to force entry into multiple openings.
   c) Heavy swinging gates, roll-down steel doors, and steel shutters. Additional staffing are required to ensure all exterior openings are forced. This must take place simultaneously with advancement of hoselines and ventilation operations.

89) **Sometimes multiple dwellings are located in densely populated congested areas. Which of the following are going to be constant considerations within your street condition size-up?**

   a) Street width, traffic congestion, and double-parked cars.
   b) Sequence of responding companies, traffic congestion, and traffic patterns.
   c) Street width, traffic patterns, and the distance between cross streets.
   d) Sequence of responding companies, overhead obstructions, and traffic pattern.

90) **The most problematic building system in high-rise buildings for firefighters responding to modern high-rise buildings is the:**

   a) Automatic sprinkler and standpipe system.
   b) Building HVAC system.
   c) Building elevators.
   d) Building fire pump.

Final Exam Questions

91) **When no life is in danger and the fire in a vacant building is threatening both exposures, chose the correct size-up points that will help identify the most severely exposed building:**

   a) Exposure proximities, exterior sheathing, and fire load of the exposure.
   b) Exposure proximities, height and area of the exposures, and the value of the content in the exposure building.
   c) Exposure proximities, exterior sheathing, and auxiliary appliances in the exposure buildings.
   d) Both **a** and **b** are correct.

92) **If windows must be opened or broken during a fire in a high-rise building, this must be accomplished under the direct control of the incident commander and the operations officer because this will:**

   a) Enable the occupants above the fire to remain in place if the windows on the fire floor are removed or opened.
   b) Make fire attack easier if the windows are opened on the downwind side of the building, keeping fresh air at the back of the advancing hose crew.
   c) Rain down large, heavy, lethal pieces of glass and may effect fire and smoke conditions in the building.
   d) Windows on the upper floors of high-rise buildings should never be opened or broken as you can never predict the wind speed or direction at the upper stories of the building.

93) **A firefighter safety concern for operating in high-ceiling warehouse occupancies is:**

   a) The danger associated with falling from a mezzanine in the building.
   b) Fire conditions above at ceiling or roof level may be drastically different from the conditions encountered at floor levels.
   c) The danger of objects such as light fixtures and unit heaters falling from the high ceiling.
   d) Water from discharging sprinkler heads may soften the bottom of pallet loads of stock, causing the unstable piles to fall on firefighters or block their path of exit.

94) **Key factors to making the strategic decision on whether to launch an offensive or defensive attack at a church fire are:**

a) Identifying the construction type and the location of the fire.
b) The proximity and construction features of the exposure buildings.
c) The ability to obtain immediate information on the fire's location and extent.
d) The ability to force entry quickly and effect ventilation so hoselines can be quickly advanced to the seat of the fire.

95) **Forcible entry concerns for both row frame and brownstone structures follow the forcible entry options at most residential structures with the addition of the following:**

a) An ornamental steel security gate may be present on the front door.
b) Roll-down security gates may be installed on the rear windows and doors on the basement and front floor levels.
c) Security gates may be installed over the windows at the basement or 1st floor level.
d) Both answers **a** and **b** are correct.

96) **Buildings built on a sloped grade cause an increased life hazard concern to firefighters because:**

a) A building that is 2 stories in the front may, in fact, be 3 or 4 stories from the rear; this may cause confusion and make fireground management more difficult.
b) Steep slopes to grade will always preclude deployment of ground ladders to upper stories, placing firefighters searching the upper stories of the building at risk.
c) Stories that open to a different grade at the rear of the building may confound hoseline deployment by confusing which level is the 1st floor of the structure; this will make fireground management more difficult.
d) Steep slopes to grade will make the use of aerial ladders more difficult and dangerous to firefighters.

Final Exam Questions

97) Which of the following statements is true regarding the unenclosed interior stairway in a private dwelling?

    a) The interior stairway is not a major concern as most fires start on the 2nd or 3rd floor of the private dwelling.
    b) This stairway will allow unrestricted movement of fire and smoke immediately to the upper floors of the private dwelling.
    c) The backup hoseline must always be stretched to the upper floors of private dwellings to prevent fire extension to the upper floors.
    d) The interior stairway is only a major concern when the fire is located in the basement or below-grade levels.

98) The stone treads and landings used in the stairways of many multiple dwellings with "V" or "U" return stair designs is a concern for firefighters because:

    a) Water runoff from firefighting operations may collect on the stair landing. The weight of the water and the firefighters may cause the landing to collapse.
    b) The sudden heating and cooling of the stone may affect its integrity, possibly causing it to collapse under the weight of a firefighter.
    c) The stone treads and landings tend to crumble as the result of exposure to heat. This will delay the advancement of hoselines.
    d) The use of stone treads and landings in buildings is of no concern to firefighters because these materials are non-combustible and will not be affected by the fire or firefighting operations.

99) Barring the use of pre-incident information, obtaining initial information about the size of the building and any irregular shapes associated with taxpayer buildings should come from:

    a) The 2nd arriving chief responding to the rear of the building to obtain this information.
    b) From the 1st company on the roof position.
    c) Having a firefighter/fire officer walk the perimeter of the involved building or area to determine size and shape of the structure.
    d) Both **b** and **c** are correct answers.

100) Which of the following are common construction concerns for fire officers operating at a fire in garden apartments and townhouse structures?

a) Unit separation voids, lightweight or hybrid construction methods and materials, vertically stacked pipe chases, and back-to-back kitchens.
b) Unit separation voids, lightweight or hybrid construction methods and materials, the height of the buildings, and back-to-back kitchens.
c) Unit separation voids, the location of garages under the buildings, vertically stacked pipe chases, and back-to-back kitchens.
d) Unit separation voids, lightweight or hybrid construction methods and materials, the location of firewalls or fire separation walls, and the height of the building.

Fireground Size-Up Study Guide

FINAL EXAM ANSWER KEY

Question #	Answer	Page #	Chapter #
1	b	1	1
2	c	6	1
3	c	145, 146	4
4	d	199	5
5	c	199, 200	5
6	a	244	7
7	d	55	2
8	c	325	9
9	d	280	8
10	b	355	10
11	d	113	3
12	b	213	6
13	b	107	3
14	a	52	2
15	b	168	4
16	a	283	8
17	a	375	10
18	c	235	6
19	c	246	7
20	a	347	9
21	c	16	1
22	d	170	4
23	a	232	6
24	b	344	9
25	c	77	2
26	d	206	5
27	a	281	8
28	c	371	10
29	d	105	3
30	b	264, 265	7
31	a	368	10
32	c	254, 255	7
33	a	106	3
34	c	41	1
35	d	162	4
36	b	233	6

37	b	343	9
38	c	67	2
39	b	201	5
40	d	289	8
41	c	358	10
42	d	22	1
43	d	118	3
44	b	277	8
45	c	244	7
46	a	83	2
47	b	159	4
48	d	223	6
49	b	323, 333	9
50	d	198	5
51	b	5/6	1
52	b	157	4
53	a	265	7
54	a	333	9
55	b	361	10
56	c	56	3
57	d	86	2
58	c	190	5
59	a	286	8
60	d	222	6
61	d	262	7
62	d	362	10
63	b	145	4
64	c	24	1
65	b	136	3
66	b	219	6
67	c	275	8
68	c	82	2
69	d	318	9
70	b	190	5
71	c	217	6
72	d	39, 40	1
73	b	140	3
74	d	362	10
75	c	279	8
76	b	52	2
77	c	184	5

78	c	317	9
79	a	156, 157	4
80	c	259	7
81	d	32, 33	1
82	a	253	7
83	d	366	10
84	b	78	2
85	c	195	5
86	a	218	6
87	d	173	4
88	c	282	8
89	a	120	3
90	b	315	9
91	d	374	10
92	c	326	9
93	b	285	8
94	c	248	7
95	c	220, 221	6
96	a	26	1
97	b	79	2
98	b	110	3
99	d	166	4
100	a	180, 183, 206	5